Industry 4.0 and Intelligent Manufacturing Technologies

The world is undergoing a transformation driven by Industry 4.0 and intelligent manufacturing technologies. This book explores aspects of well-being and collaboration in industrial systems which were underscored by automation and security focus.

Industry 4.0 and Intelligent Manufacturing Technologies explores the different ways in which technology disrupts and revolutionizes industrial management. It discusses the connection between engineering, science, and management, aiming to assist companies, researchers, and professionals navigate and excel in this transformation era. This book offers insights by examining Industry 4.0 and intelligent manufacturing technologies. It looks into the driving forces behind digitalization, automation, intelligent process control, artificial intelligence, and additive manufacturing. Lastly, the book showcases the seamless integration of advanced digital technologies into manufacturing and industrial processes.

This book aims to attract managers and researchers to the field of industrial engineering. It also offers valuable insights for students, graduates, and educators focusing on industrial and business management. However, the content is not exclusive to these audiences and will be engaging for any reader with an interest in the subject matter.

Industrial and Systems Engineering Series
Series Editors: Waldemar Karwowski, *University of Central Florida, Orlando, USA* & **Hamid Parsaei,** *Texas A&M University*

Industrial Engineering has evolved as a major engineering and management discipline, the effective utilization of which has contributed to our increased standard of living through increased productivity, quality of work, and quality of services and improvements in the working environments. The Industrial and Systems Engineering book series provides timely and useful methodologies for achieving increased productivity and quality, competitiveness, globalization of business and for increasing the quality of working life in manufacturing and service industries. This book series should be of value to all industrial engineers and managers, whether they are in profit-motivated operations or in other nonprofit fields of activity.

Managing Professional Service Delivery: 9 Rules for Success
Barry M. Mundt, Francis J. Smith, and Stephen D. Egan Jr.

Mobile Electronic Commerce: Foundations, Development, and Applications
Edited by June Wei

Manufacturing Productivity in China
Edited by Li Zheng, Simin Huang, and Zhihai Zhang

Human Factors in Transportation: Social and Technological Evolution Across Maritime, Road, Rail, and Aviation Domains
Giuseppe Di Bucchianico, Andrea Vallicelli, Neville A. Stanton, and Steven J. Landry

Laser and Photonic Systems: Design and Integration
Edited by Shimon Y. Nof, Andrew M. Weiner, and Gary J. Cheng

Supply Chain Management and Logistics: Innovative Strategies and Practical Solutions
Edited by Zhe Liang, Wanpracha Art Chaovalitwongse, and Leyuan Shi

Production Management, Manufacturing, and Process Control
Edited by Beata Mrugalska, Waldemar Karwowski, and Tareq Z. Ahram

Industry 4.0 and Intelligent Manufacturing Technologies
Edited by Beata Mrugalska, Waldemar Karwowski, and Tareq Z. Ahram

For more inform ation about this series, please visit: www.routledge.com/Industrial-and-Systems-Engineering-Series/book-series/CRCINDSYSENG

Industry 4.0 and Intelligent Manufacturing Technologies

Edited by
Beata Mrugalska, Waldemar Karwowski, and
Tareq Z. Ahram

Routledge
Taylor & Francis Group
NEW YORK AND LONDON

Designed cover image: Shutterstock

First published 2026
by Routledge
605 Third Avenue, New York, NY 10158

and by Routledge
4 Park Square, Milton Park, Abingdon, Oxon, OX14 4RN

Routledge is an imprint of the Taylor & Francis Group, an informa business

ISBN: 978-1-041-03714-9 (hbk)
ISBN: 978-1-041-03788-0 (pbk)
ISBN: 978-1-003-62536-0 (ebk)

DOI: 10.1201/9781003625360

Typeset in Times
by Newgen Publishing UK

Contents

About the Editors

Beata Mrugalska, Ph.D., D.Sc., Eur.Erg., is Associate Professor, Vice Dean for Science (2024–2028) at the Faculty of Management Engineering, and Head of the Division of Applied Ergonomics (2020–2024) (2024–2028), Institute of Safety and Quality Engineering, Faculty of Management Engineering, Poznan University of Technology in Poland. She holds an M.Sc. degree (2001) in Management and Marketing from the Faculty of Mechanical Engineering and Management at the Poznan University of Technology and a Ph.D. (2009) in Machine Construction and Operation from the Faculty of Computer Science and Management at the Poznan University of Technology. She was awarded D.Sc. (dr habil.) degree in Mechanical Engineering by the Faculty of Mechanical Engineering of the Poznan University of Technology, Poland (2019). Since 2018, she has been a board member of the Center for Registration of European Ergonomists (CREE). She is responsible for promoting the professional title of EuroErgonomist in the international scientific and industrial environment. In 2024 she was elected to the board of Association of Engineering, Project, and Production Management (EPPM) for the 2024–2026 term. Dr. Mrugalska is also a board member of the Polish Academy of Arts and Sciences, Commission of Ergonomics, and a member of the Ergonomics Committee at the Polish Academy of Sciences, Branch Poznan. She has over 150 publications on human factors in modern organizational management concepts. She serves as a member of eight editorial boards of international journals and a guest editor of five special issues of journals. Dr. Mrugalska was honored with the top peer-reviewer by Publons for her contribution to the preparation of reviews of papers.

Waldemar Karwowski, Ph.D., D.Sc., P.E., is Pegasus Professor and Chairman, Department of Industrial Engineering and Management Systems, and Executive Director, Institute for Advanced Systems Engineering, University of Central Florida, Orlando, USA. He holds an M.S. degree in Production Engineering and Management from the Technical University of Wroclaw, Poland, and a Ph.D. in Industrial Engineering from Texas Tech University. He was awarded a D.Sc. (dr habil.) degree in management science by the State Institute for Organization and Management in Industry, Poland. Dr. Karwowski is the past President of the International Ergonomics Association (2000–2003) and the past President of the Human Factors and Ergonomics Society (2007). He served on the Committee on Human Factors/Human Systems Integration, National Research Council, and the National Academies, USA (2007–2011). He has over 600 publications focused on human performance, safety, cognitive systems engineering, human-centered design, neuro-fuzzy systems, nonlinear dynamics, neurotechnology, and neuroergonomics. Dr. Karwowski serves as Co-Editor-in-Chief of Theoretical Issues in Ergonomics Science journal (Taylor & Francis, Ltd), Editor-in-Chief of Human-Intelligent Systems Integration (Springer), and Field Chief Editor of Frontiers in Neuroergonomics.

Tareq Z. Ahram, Ph.D., is a Senior Executive and Research Professor in Emerging Technologies and Intelligent Healthcare Systems in the USA and a Visiting Professor in London, UK. He received a Ph.D. degree in Industrial and Systems Engineering from the University of Central Florida, USA, in 2008, with specialization in human systems integration and large-scale information retrieval systems optimization; an M.Sc. in Industrial Systems Engineering with a focus on Human Factors from UCF in 2007, and an M.Sc. and B.Sc. in Industrial and Systems Engineering from the University of Jordan in 2002 and 2004 respectively.

Contributors

Alireza Abouhossein
Department of Ergonomics, School of Public Health and Safety, Shahid Beheshti University of Medical Sciences
Tehran, Iran

Jami Aho
Turku University of Applied Sciences Ltd
Turku, Finland

Joseph Anderson
Institute for Process Excellence (IpX)
Denver, CO, USA

Zahra Ansari
Department of Ergonomics, School of Public Health and Safety, Shahid Beheshti University of Medical Sciences
Tehran, Iran

Susanna Aromaa
VTT Technical Research Centre of Finland Ltd
Espoo, Finland

Sanderson Cesar Macedo Barbalho
Faculty of Technology, University of Brasilia
Distrito Federal, Brazil

Anja Braun
NXT Sustainability and Technology, Reutlingen University
Reutlingen, Germany

Jorge D. Camba
Purdue University
West Lafayette, IN, USA

Artur Costa
University of Minho
Azurém, Guimarães, Portugal

Tihomir Dovramadjiev
Technical University of Varna
Varna, Bulgaria

Travis Fuerst
Purdue University
West Lafayette, IN, USA

Mariannys Rodríguez Gasca
Faculty of Technology, University of Brasilia
Distrito Federal, Brazil

Gabriele Goretti
DESTEC Department, University of Pisa
Pisa, Italy

Antero Karvonen
VTT Technical Research Centre of Finland Ltd
Espoo, Finland

Hanna Koskinen
VTT Technical Research Centre of Finland Ltd
Espoo, Finland

Mika Luimula
Turku University of Applied Sciences Ltd
Turku, Finland

Jeffrey Mahachi
School of Civil Engineering and the Built Environment, University of Johannesburg
Johannesburg, South Africa

Georgios Markopoulos
Department of Management, London School of Economics and Political Science
London, England

Panagiotis Markopoulos
London Digital Institute, Staffordshire University
London, England

Beata Mrugalska
Faculty of Engineering Management, Poznan University of Technology
Poznan, Poland

Marcin Mrugalski
Institute of Control and Computation Engineering, University of Zielona Gora,
Zielona Gora, Poland

Mari Myllylä
Faculty of Information Technology, University of Jyväskylä
Jyväskylä, Finland

Akash Nandi
Department of Economics, Harvard University
Cambridge, United States of America

Peter Ohlhausen
NXT Sustainability and Technology, Reutlingen University,
Reutlingen, Germany
Fraunhofer-Institut für Arbeitswirtschaft und Organisation IAO
Stuttgart, Germany

António J. Pontes
University of Minho
Azurém, Guimarães, Portugal

Sajjad Rostamzadeh
Department of Ergonomics, School of Public Health and Safety, Shahid Beheshti University of Medical Sciences
Tehran, Iran

Nada Ruzicic
NXT Sustainability and Technology, Reutlingen University
Reutlingen, Germany

Pertti Saariluoma
Faculty of Information Technology, University of Jyväskylä
Jyväskylä, Finland

Mascha Samkova
Sciences Po, Paris School of International Affairs
Paris, France

Álvaro M. Sampaio
University of Minho
Azurém, Guimarães, Portugal

Afonso Santos
University of Minho
Azurém, Guimarães, Portugal

Eugenia Cornils Monteiro da Silva
Faculty of Technology, University of Brasilia
Distrito Federal, Brazil

Antti Tammela
VTT Technical Research Centre of Finland Ltd
Espoo, Finland

Bastian Tammentie
VTT Technical Research Centre of Finland Ltd
Espoo, Finland

Benedetta Terenzi
DESTEC Department, University of Pisa
Pisa, Italy

Preface

The world is undergoing a global transformation driven by the implementation of Industry 4.0, intelligent manufacturing technologies, and artificial intelligence. This book explores the shift from emphasizing automation and security to employees' well-being and human-AI collaboration in contemporary industrial systems.

Industry 4.0 and Intelligent Manufacturing Technologies explores how technology disrupts and revolutionizes contemporary industrial management. The book explores the intersection of engineering, science, and management, aiming to help companies, researchers, and professionals navigate and excel in this transformative era This book also offers insights by examining Industry 4.0 and intelligent manufacturing technologies. It looks into the driving forces behind digitalization, automation, intelligent process control, artificial intelligence, and additive manufacturing. Lastly, the book showcases the seamless integration of advanced digital technologies with manufacturing and industrial processes.

The book features carefully selected chapters from the papers presented at the Applied Human Factors and Ergonomics (AHFE) Conference, with a primary focus on production management and control. This track explores the exchange of ideas and advancements in production, sustainability, life cycle management, innovation, fault diagnostics, and control systems. It provides an excellent forum for exploring the interconnection and collaboration between researchers and practitioners from academia and industry.

Chapter 1 deals with the importance of change and configuration management in achieving operational excellence in industrial enterprises, particularly in the context of Industry 4.0 and smart manufacturing. Despite their importance, they are often overlooked from an educational standpoint. Thus, this chapter outlines a vision for a digital enterprise curriculum featuring two courses developed in collaboration with industry. These courses, aimed at graduate and undergraduate students, focus on change and configuration management using the industry-standard CM2 methodology. They include practical exercises to connect theoretical principles with the supporting software technology.

Chapter 2 examines the impact of Industry 4.0 technologies on manufacturing and work dynamics through a comprehensive literature review and bibliometric analysis. There has been a notable increase in interest in human factors and ergonomics related to these technologies, with a significant surge since 2019, partially due to COVID-19's influence on operational changes. As Industry 5.0 emerges, human factors have become a central focus in industrial processes.

Chapter 3 discusses the increasing use of automated guided vehicles (AGVs) in Industry 4.0, highlighting their benefits, including high mobility, reliability, cost-effectiveness, and environmental advantages. Despite these benefits, AGVs face scheduling and control challenges due to uncertainties in the transport system. To address this issue, a novel centralized AGV scheduling algorithm is proposed, utilizing the model predictive control (MPC) method. This approach utilizes max-plus linear algebra (MPLA) to develop an efficient real-time work plan, making the algorithm applicable to various transportation systems in Industry 4.0. Chapter 4 addresses the challenges of transitioning from linear to circular value creation systems, which often weren't initially designed with circularity in mind. It introduces an interdependency matrix to aid in understanding and integrating circularity into value systems. It enhances the original matrix with practical examples and further explanations using a mixed-methods approach. The interdependency matrix helps evaluate the relationships between various influence factors, providing a holistic view and identifying independent and context-dependent factors within a CE system.

Chapter 5 refers to the "Made in Italy" brand, which exemplifies the integration of innovation and tradition, enhanced by European support for digitalization. Industry 5.0 is a framework for optimizing efficiency while preserving core values, emphasizing human creativity, synergy, sustainability, and personalized production.

Chapter 6 explores the potential of smart cities to positively impact the environment and society by promoting sustainability, enhancing accessibility, reducing emissions, and fostering economic growth and climate action. It examines the role of metaverse technologies as experimental grounds for smart city concepts, providing insights for real-world adaptation. It highlights key technologies for digital ecosystems, analyzes user engagement, and aligns these with the UN Sustainable Development Goals (SDGs). Furthermore, the chapter proposes a framework to identify the necessary characteristics and technological requirements for smart cities, thereby facilitating their development in line with the SDGs.

Chapter 7 illustrates the transformation of industrial process control, driven by technological advancements such as intelligent technologies and the industrial metaverse, alongside societal demands for sustainability and human-centric approaches. It introduces a framework and methodology for developing a human-centric, interdisciplinary, and collaborative research agenda. It emphasizes the importance of integrating specific research activities with the overall agenda, supported by empirical findings from case studies on intelligent technologies, AI assistants, and the industrial metaverse.

Chapter 8 explores the advancements in additive manufacturing using robotic arm extrusion and its application in creating large-scale solutions. It emphasizes the importance of adopting new approaches in design, material selection, and production control to fully leverage these technologies. A primary focus is on outdoor urban green spaces, where such manufacturing techniques can enhance social interactions through innovative products.

Chapter 9 discusses recent advancements in user-centered design (UCD), which is facilitated by technologies such as additive manufacturing, artificial intelligence, and incorporating anthropometry and biomechanics. These advancements have allowed for the creation of tailored products, systems, and environments. Industry 4.0 and Industry 5.0 have further integrated UCD, enhancing productivity and design innovation. In this chapter, two UCD processes are highlighted: one for a hands-free door handle to prevent pathogen transfer in hospitals and another for a crochet hook handle to improve ergonomic comfort and reduce pain associated with crocheting. Both projects refer to anthropometric data and subjective assessment tools.

Chapter 10 explores the potential of modular construction (MMC) and 3D printing in addressing South Africa's affordable housing crisis, with a focus on sustainable urbanization. It notes that while the National Building Regulations allow these modern construction methods, they must meet rigorous standards set by Agrément South Africa, including structural and performance assessments. The chapter aims to establish a comprehensive knowledge base on MMC and 3D-printed materials, examining their role in sustainable urban development. It identifies challenges in deploying these certified technologies, particularly in affordable housing, and proposes a framework that integrates research, innovation, government support, and market strategies to enhance their adoption in the housing sector.

Chapter 11 explores how innovations can enhance the inclusivity of healthcare in the digital era. To achieve it, it suggests focusing on broad meta-levels of education that should be incorporated into higher education curricula. The proposed course of action is designed to fill the gap by offering a student-focused curriculum that combines theoretical knowledge with hands-on activities, aiming to improve digital skills and boost student employability.

This book is designed to captivate managers and researchers in industrial engineering. It also offers valuable insights for students, graduates, and educators focusing on industrial and business management. However, the content is not exclusive to this specific audience; rather, it engages readers interested in multiple subject matters.

The Editors
Beata Mrugalska
Waldemar Karwowski
Tareq Z. Ahram

Acknowledgements

The editors would like to thank all the authors who contributed to this work. We also wish to thank the following scientific and advisory board members for their diligence and expertise in carefully selecting and reviewing the papers:

Salvador Ávila Filho (Federal University of Bahia), Brazil
Murray Gibson (Saturn Ergonomics Consulting), USA
Akihiko Goto (Osaka Sangyo University), Japan
Hiroyuki Hamada (Kyoto Institute of Technology), Japan
Adam Hamrol (Poznan University of Technology), Poland
Aidé Aracely Maldonado Macías (Universidad Autónoma de Cd. Juárez), Mexico
Jerry Poh Kiat Ng (Multimedia University), Malaysia
Jörg Niemann (University of Applied Sciences), Germany
Edgar Ramos (Universidad Peruana de Ciencias Aplicadas (UPC), Peru
Carlos Raymundo (Universidad Peruana de Ciencias Aplicadas (UPC), Peru
Silvio Simani (University of Ferrara), Italy
Yusuf Tansel İç (Başkent Üniversitesi), Turkey
Benedict Wohlers (Fraunhofer IEM), Germany
Magdalena Wyrwicka (Poznan University of Technology), Poland

1 Change and Configuration Management in the Digital Enterprise Curriculum

Jorge D. Camba, Travis Fuerst, and Joseph Anderson

INTRODUCTION AND BACKGROUND

Computer information and communication technologies have been essential companions for engineering organizations for decades. Today, virtually any process, activity, and operation across an enterprise relies on or is driven by digital technology to the point that it would be impossible to imagine any engineering discipline without this connection to the software world. And the connection is ever-strengthening.

As more companies embark on digital transformation initiatives and revamp their current practices, operations, and ways of working, concepts such as digital twin and digital thread have become critical vehicles to harness the power of automation, data analytics, artificial intelligence (AI), cloud computing, and other emerging digital technologies. In these environments, the "digital product," that is, the computer representation of the system or artifact, becomes a key information asset around which all development activities revolve (Aranburu et al., 2022a).

The goal of a digital transformation is to enable smarter decision-making and effective real-time response to market disruptions while reducing costs and improving quality (Jones et al., 2021). In this context, the model-based enterprise (MBE) paradigm and enterprise software systems such as product lifecycle management (PLM), manufacturing execution system (MES), and enterprise resource planning (ERP) are considered major drivers of digital innovation and progress and provide the digital backbone for the Fourth Industrial Revolution (Philbeck and Davis, 2018; Gromova, 2020). A lack of knowledge of these powerful yet complex systems can certainly lead to significant challenges for engineering organizations. However, evidence shows that many of the problems and negative experiences attributed to technology are symptoms of an underlying bigger issue, which often stems from inefficient business

DOI: 10.1201/9781003625360-1

processes and workflows, especially those related to the management of changes (Nyemba et al., 2021; Eckert and Clarkson, 2023).

As engineering projects become larger and more complex and involve geographically dispersed teams, sometimes speaking different languages, the nature of the communication, and collaboration efforts required to successfully execute the project drastically change. The complexity of the challenges in terms of change and configuration management also increases. Additionally, many organizations spend most of their time working with existing systems, either changing, adapting, improving, or reusing them, rather than starting new projects entirely from scratch. In this regard, change management activities can easily become as burdensome as developing and supporting customers' products and solutions, even when powerful enterprise technology systems are used. Developing, maintaining, and making changes in large engineering systems over time are extremely difficult (Dart, 1992). In fact, many organizations continue to struggle to define a fast and efficient change management process. Many organizations have changed or replaced their change process multiple times without understanding the dynamics of change or the building blocks needed to facilitate change management.

In smaller and younger organizations, the situation is not any better. Although the scale of the projects is often smaller, change and configuration management is typically not enforced, which contributes to a subtle but relevant effect known as "complexity creep," the trend of a product, project, system, or process to become increasingly complex over time (Chechile, 2021).

From an intervention standpoint, Arthur (1988) considered three types of actions to be performed on a project. The classification was presented in the context of software maintenance, but it can be generalized to any engineering system:

1. *Corrective Action*: Rework activities and effort required to correct and/or compensate for something that should not have happened in the first place. Any time spent in a state of uncertainty about what to do, or how to do it, is also a form of corrective action. Corrective action is unplanned and costly. It translates to intervention resource expenditures and ultimately represents waste.
2. *Adaptive Action*: Change required because some improvement or enhancement must be applied to a system, such as the addition of new hardware.
3. *Perfective Action*: Change that is performed to make the system more flexible, robust, and responsive to future changes. Perfective action drives quality, continuous improvement, and innovation.

In environments characterized by inefficient business processes and workflows, intervention resource expenditures – the resources expended to compensate for quality and schedule problems – are notoriously high, costly recalls are often factored into warranty costs, and corrective action becomes the standard "way of working." More and more firms realize that most change problems are related to complicated and congested administrative processes, and many organizations are exploring best practices to fit with their business scenarios (Jarratt et al., 2011).

In the software realm, effective change and configuration management practices have been consistently identified as critical process-related factors that determine project success (Oz and Sosik, 2000; Schmidt et al., 2001; Nasir and Sahibuddin, 2011). Likewise, in the dynamic landscape of modern engineering and manufacturing, the ability to implement robust yet flexible change processes that minimize corrective action and maximize perfective action is not just a competitive advantage but a necessity. Indeed, leading manufacturing companies agree that configuration management has become more, rather than less, important in complex engineering (Lindkvist et al., 2013). Furthermore, some researchers have recently discussed the emergence of the "configuration manager" as a new professional figure, that is, a cross-functional expert on the development of mass customization strategies and product variety management (Sandrin et al., 2022).

Change and configuration management processes provide a standardized methodology to alter the product configuration in a rigorous and procedural manner and ensure that a product's definition and its realization are in alignment. In the words of Chechile (2021), the purpose of configuration management is "to establish and maintain consistency of the system's performance, functional and physical attributes with its requirements, design and operational information throughout its life."

Change and configuration management as a discipline can be formally traced back to the industrial surge following World War II, heavily influenced by the military and space advancements driven by the Cold War. At that time, there was (and continues to be) a need to ensure that the swift development of technology and products was effectively managed to support the design, production, and integration of ongoing research and development activities. Over time, change and configuration management transitioned from a Department of Defense technical management process focused solely on physical items to a cross-functional standard utilized across various industries. Most notably, as software became a more prevalent component of engineering systems and IT infrastructures, the scope of change and configuration management expanded to include both hardware and software.

The introduction of version control systems such as source code control system (SCCS) and revision control system (RCS) represents a significant milestone that enabled organizations to control changes to source code. Likewise, the advent of structured programming methodologies and subsequent client–server architectures further highlighted the need and challenges of systematic configuration control. The demand for more advanced mechanisms became apparent, which set the stage for the development of comprehensive frameworks, such as the software configuration management (SCM) process (Estublier et al., 2005; Whyte et al., 2016; Farayola et al., 2023).

The evolution of change and configuration management as a formal discipline also brought changes in the specifications. The multiple military standards (480 series) that governed configuration management until 1992 were merged into a single specification, MIL-STD-973 (USDoD, 1992). This specification later evolved through a collaborative effort between government and industry, eventually resulting in ANSI/EIA-649 (American National Standards Institute/Electronic Industries Alliance)

(SAE, 2019), the National Consensus Standard for Configuration Management. According to the ANSI/EIA-649C standard, configuration management involves "applying appropriate processes, resources, and controls, to establish and maintain consistency between product configuration information, and the product" (SAE, 2019). A configuration management standard that builds on the ANSI/EIA-649 standards and is widely adopted in both industry and defense is CM2 (IpX, 2024). This model enhances the configuration management process by broadening its scope across the enterprise and incorporating process excellence. CM2 is issued by the Institute for Process Excellence (IpX).

Despite its relatively long history and critical importance in the enterprise, change and configuration management continues to be a challenge for most organizations and a recurrent concern, especially in large engineering projects. These challenges are significant at any level of detail of a project, from properly managing item numbers in product structures and bills of materials to effectively implementing changes to the digital product model in its native computer-aided design (CAD) format (Camba et al., 2021; Aranburu et al., 2022b). The dynamic and competitive nature of modern engineering, product development, and manufacturing, with frequent updates and releases, and the constant pressure to innovate exacerbate the challenge of maintaining accurate and up-to-date configuration information (Farayola et al., 2023).

From a higher education standpoint, the traditional engineering curriculum has continuously adapted to the emergence of digital technology by incorporating more time with computers, programming classes, and the use of various software systems and applications (Morris and Wagner, 2023). However, the management and logistical aspects of engineering, in terms of the roles students will have to fulfill in modern industries and the way they are expected to operate and work as part of a team in a digital enterprise ecosystem, as well as the consideration of lifecycle engineering as a fully integrated approach to product development, are widely ignored. Enterprise systems such as PLM and topics such as business processes or change and configuration management are usually left out of the curriculum. In some cases, they are covered indirectly (e.g., in a CAD course) yet often insufficiently (due to time constraints) by instructors with a genuine interest in the subject or as an extracurricular activity (Morris and Wagner, 2023). In general, engineering students finish their studies with suboptimal skills in PLM topics, change and configuration management, lifecycle engineering, and the digital enterprise ecosystem as a whole (CIMdata, 2019).

This article builds on our previous paper presented at the International Conference on Applied Human Factors and Ergonomics (AHFE 2024) (Camba and Anderson, 2024). We describe two courses: a graduate-level change and configuration management course and an equivalent undergraduate-level course developed at Purdue University in collaboration with the IpX. IpX is the organization behind the CM2 model (IpX, 2024), the industry standard for enterprise change and configuration management used by both governmental and private enterprises worldwide. We discuss the motivation, objectives, structure, and assessment methods for the course and report some student feedback and lessons learned.

MOTIVATION

From an engineering management point of view, change and configuration management is a particularly critical element of the Industry 4.0 ecosystem, playing a vital role in ensuring the quality and traceability of products and designs. These disciplines encompass a set of principles and practices—often supported by software tools—to meticulously track the changes made to project artifacts and deliverables and control a project's history and evolution (Jokic et al., 2010). As mentioned earlier, various standards in support of change and configuration management are available and have been adopted by industry, including MIL-STD-973 (USDoD, 1992), ANSI/EIA-649C (SAE, 2019), ISO 10007:2017 (ISO, 2017), IEEE Std. 828-2012 (IEEE, 2012), and IpX's CM2 (IpX, 2024). Change and configuration management is also a key aspect of quality assurance and certifications such as ISO 9001 or Capability Maturity Model Integration (CMMI).

When we think of an organization's processes and data, we traditionally consider these elements as simple tools or steps in the system that help an engineering organization accomplish a goal. However, a corporation's data is a foundational building block that allows it to leverage institutional knowledge, optimize current product offerings, and conceptualize products with complete end-to-end traceability. IpX's CM2 model uses a DNA analogy to illustrate this idea and the interplay between systems and data vs. people and process through the lifecycle (IpX, 2024), as shown in Figure 1.1. Individually, they each tell part of the story from a different perspective and with a different focus. But connected, they build upon one another to create a product or project, allowing the organization to work initiatives and challenges tactically with a confidence that the scope of the change is understood not only at its current point in the lifecycle but with an understanding of downstream implications to ensure the decisions being made address the full lifecycle of the product. According to IpX, breakthrough business process improvement outcomes are achieved by holistically addressing the interoperability of an organization's people, processes, systems, and data. Technology enables process change, but desirable business improvement outcomes cannot be achieved without organizational changes. Ultimately, people must be inspired, empowered, and engaged.

Change and configuration management is at the core of an engineering and technology organization's processes and data, and it affects every member of the enterprise. These disciplines provide strategies for ensuring the quality, reliability, and performance of complex systems in various industries. Change and configuration management involves identifying, documenting, controlling, and tracking changes to the system's components, requirements, and documentation throughout its lifecycle. It also helps to manage the integration of these changes into the final product and verify that the system meets its intended specifications (Hamraz et al., 2013).

Many authors have discussed the challenges that engineering schools face to teach the new technical skills and technologies—many of them disruptive—required to work and thrive in the emerging Industry 4.0 ecosystem (Frame, 2004; Boyd et al., 2023). Traditional engineering curriculum emphasizes mathematics, physics, and the engineering sciences and combines them with practical and applied hands-on skills such as programming, spatial visualization, and CAD (Agost et al., 2022;

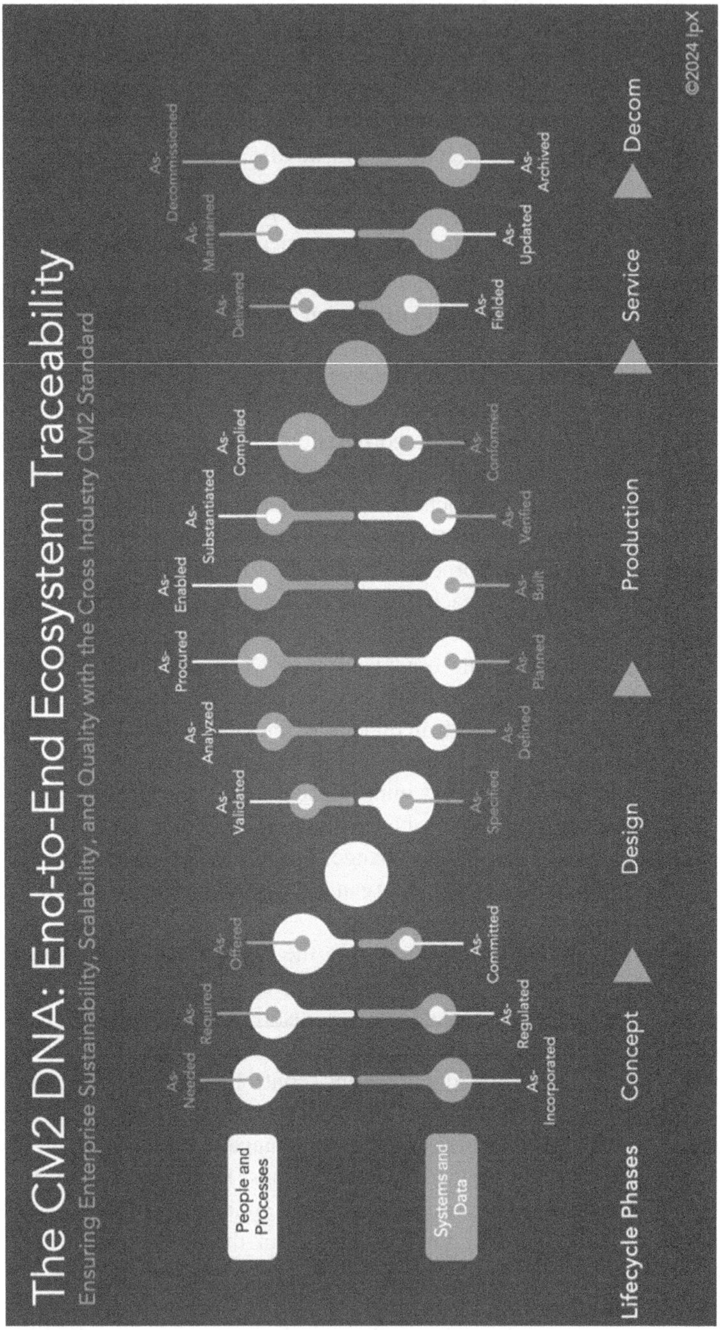

FIGURE 1.1 The CM2 DNA as illustrated by the Institute for Process Excellence. (Source: IpX, 2024)

Bartlett and Camba, 2023). The breadth and depth of knowledge, both practical and theoretical, embedded in these curricula combined with extensive accreditation requirements affect the flexibility of these programs. As a result, it is often difficult to make changes or add more courses or credits to the curriculum (Johnson et al., 2012) which is problematic as new technologies, tools, and especially processes and methods developed for Industry 4.0 are becoming increasingly prevalent and relevant to real-world industry. This challenge is critical given the growing demand for skilled engineers and technology professionals in support of Industry 4.0.

Unfortunately, despite significant investments in digital transformation initiatives by engineering and manufacturing firms, higher education has been comparatively slow to adapt to this new landscape. Certainly, engineering curricula are continuously evolving to incorporate emerging technologies such as AI and machine learning. However, operational and business aspects of the engineering profession such as PLM, enterprise systems, and business processes, such as change and configuration management, are often overlooked and expected to be learned on the job or through internships (Fuerst et al., 2024). This gap has left many companies in a difficult position, with an existing engineering workforce hesitant to adopt new technologies and a new generation of engineers lacking the basic skills needed to function effectively in an enterprise setting. A prime example of this gap is 3D CAD, as modeling courses are often taught independently from the enterprise environment and the company's documented processes. Likewise, the case of change and configuration management is perhaps even more pronounced than that of CAD, since the topic is not even present (not even indirectly) in most engineering and technology plans of study. Nevertheless, specifically in a software context, some authors have claimed that "the benefits afforded to engineering projects by configuration management practices and principles can translate easily to student and research projects within the University" (Jokic et al., 2010).

In this paper, we describe the development and the implementation of a 7-week graduate-level online course where students have the opportunity to explore, research, and discuss the challenges and practices associated with change and configuration management in the digital enterprise and how change impacts operations management and business processes. We also describe an undergraduate version of the same course, delivered in person over a period of 16 weeks, where students explore the software mechanisms and technology that support change and configuration management. The courses emphasize the CM2 model, a robust comprehensive methodology supported by the IpX for managing the configuration of a product, system, and/or service throughout its life. CM2 teaches that communication and change affect all functions of an organization by preventing corrective action, quality escapes, and recalls.

COURSE OBJECTIVES AND STRUCTURE

The goal of the courses is to provide a solid foundation for enterprise change management based on the CM2 model. The specific learning objectives for the course include the following:

- Summarize the challenges that engineering and manufacturing organizations face when managing quality, options, and variants throughout a product lifecycle.
- Construct and justify a product hierarchy and its associated datasets.
- Estimate the magnitude of corrective action for a given change and compare it to the application of conformance measurements.
- Assess the impact of an engineering change.
- Evaluate and interpret the CM2 model, an industry-standard closed-loop change management process.
- Differentiate and prepare the main change objects for authorizing and managing work.
- Summarize the challenges and complexities of lifecycle sustainment and assess how they differ from those in product development.
- Prepare a detailed change implementation plan.

The graduate-level course is designed as a 7-week online course. The delivery method is asynchronous, where students can complete the modules in their own time. The course is structured into six modules. Each week, one full module is covered along with its respective activities and assessments. The last week of the course is reserved for reviewing the content presented in previous modules and examining and reflecting on the CM2 model as a methodology for enterprise change. In the undergraduate version of the course, each module is delivered over a period of 2 weeks. The six modules are described in Table 1.1.

The content of each module relies on reading assignments from IpX's official textbook for the CM2-Comprehensive Certification (Camba et al., 2023). After the successful completion of the course, students will be prepared to take the CM2-C certification exam. CM2 certifications have been provided worldwide since 1986 to companies of all sizes and industries making it known as the global standard for enterprise change and configuration management. The set of CM2 certifications is illustrated in Figure 1.2.

The CM2-C certification is a six-course series that provides a comprehensive methodology for managing the configuration of a product, system, and/or service throughout its life. The certification supports the various roles involved with change and configuration management in the enterprise, such as project, product, and process managers/engineers; change specialists; PLM analysts; and solutions architects. The CM2-C certification provides an expansive, enterprise-wide awareness and skillset in understanding an organization's full change process from ideation through decommission while understanding how it affects the bottom line. The CM2-C certification provides additional credentials to the students' resumes to set them apart in the hiring process.

COURSE ASSESSMENT

Course assessment is comprised of three elements: short quizzes, discussions, and assignments. Each weekly module contains these elements. Weekly quizzes are intended to evaluate the student's understanding of the reading material. All quizzes

TABLE 1.1
Module descriptions

Module	Description	Key concepts
1. Foundations of operational excellence	The first module lays the foundation for operational excellence through the CM2 industry standard by discussing the concept of business requirements and the business infrastructure that is built on this notion. How an organization identifies, structures, links, and assigns ownership to its requirements and internal processes directly affects its ability to successfully and efficiently perform the intended mission or achieve its business objectives. The role and importance of communication are emphasized as well as the concept, scope, and magnitude of corrective action and intervention resource expenditures. These expenditures are the unplanned time, money, and resources expended to compensate for quality and schedule problems	• Requirements • Communication • Conformance • CM2 business process infrastructure • Intervention resources • Corrective action
2. Baselines and the development process	The module discusses the key building blocks for change process improvement, emphasizing baselines as the mechanisms to ensure that the information being changed is properly identified, structured, linked, and owned. It also examines the CM2 approach to product development and its foundational element, the V-model. The model is grounded on the V-model of systems engineering, a well-known representation of a system's development lifecycle that directly associates development requirements with tangible deliverables	• CM2 baselines • V-model • 4-tier development process • Application requirements • Design basis • Product hierarchy • Work breakdown structure (WBS) • Detailed design
3. Items and datasets	The module examines the role, significance, and proper management of items and datasets as information representation mechanisms, emphasizing the different relationships between these elements and how they should be structured and made visible to the organization. Common strategies for naming, numbering, and reuse are explored as well as the mechanisms to ensure data integrity, validation, and proper item re-identification. Lot control and traceability of items are also discussed	• Items and datasets • Naming, numbering, and attributes • Re-identification • Ownership and approvals • Data integrity • Options and variants

(continued)

TABLE 1.1 (Continued)
Module descriptions

Module	Description	Key concepts
4. Change management foundations	The module studies the information elements, artifacts, and mechanisms that support the change management process: investigation request (IR), change request (CR), change notice (CN), application order (AO), and deviation and waiver (D/W). It discusses ways to ensure all issues of change, from a problem in innovation to a quote, are properly captured, documented, recorded, and incorporated throughout the organization	• Triggers • Control points • Change objects • Cost of change • Impact matrix • Change records • Change visibility • Effectivity
5. The change management process	Most organizations have subject matter experts identified for each key discipline, but rarely do organizations have a subject matter expert focused specifically on the management of change. This module introduces the CM2 change management process and discusses the roles (change leader, change owner, change review board, change implementation leader, change implementation board, and audit and release analyst), responsibilities, and workflows required for an organization to manage change efficiently and effectively	• CM2 change process • CM2 roles • Enterprise change assessment (ECA) • Fast-track vs. full-track changes • Grouping • Planning • Revision levels • Changes to changes
6. The complexities of lifecycle sustainment	The journey of developing a product does not end when the end item is delivered to the customer. Rather, the product moves to a new stage of its lifecycle where maintenance, support, and assistance must be provided. This module studies how to manage the information of a product once it is in service or with a customer. It examines the gaps that occur between product development, factories, and service organizations. The module also introduces critical additions to the process flows and roles previously defined in the CM2 process, which are essential to the management of the digital thread in the integrated logistics support (ILS) lifecycle phase, which pertains to operation and maintenance	• ILS terminology • ILS planning • Supply chain management for ILS • Additional product identifiers • Asset configurations • Overhaul, refurbish, and recertify • Change management in ILS • Facilities

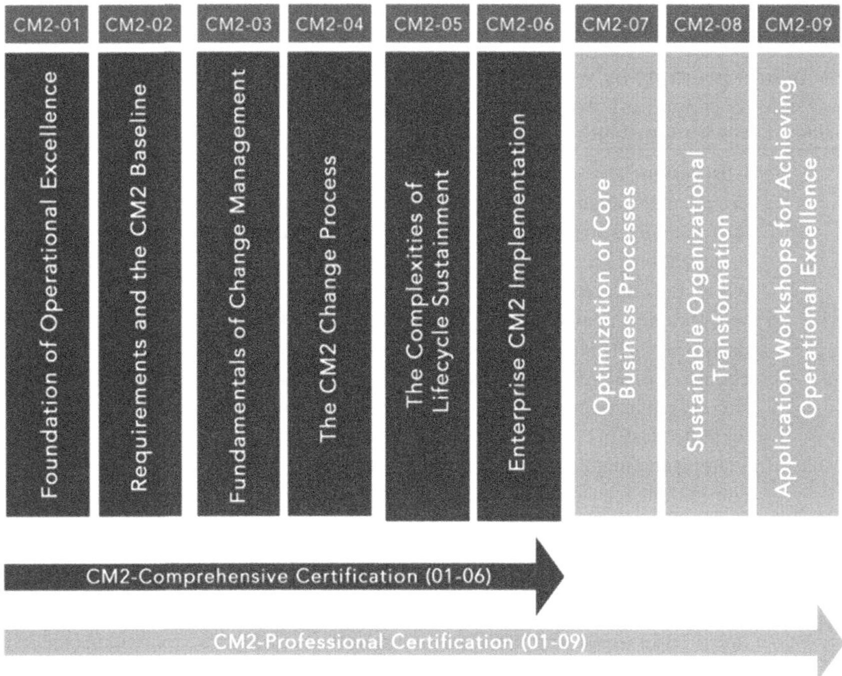

FIGURE 1.2 CM2 certification courses. (Source: https://ipxhq.com/training/cm2-certification-courses/abstracts)

include ten multiple-choice questions and are open book but cannot be taken multiple times. Class discussions are designed to promote student interactions and foster conversations related to various change management topics. Discussion prompts are posted weekly in the forum of the course delivery system. The prompts used in the course are shown in Table 1.2.

Discussion prompts serve as starting points for threads where students participate and engage with others by posting their responses, sharing comments and reflections, and comparing their own experiences to the topic. Discussion topics range from case studies of improper management of requirements and/or changes that have resulted in catastrophic failures to sharing personal experiences, anecdotes, recent news, and job announcements related to change and configuration management. Students are required to create an initial post between 200 and 400 words. To encourage discussions and student interactions, they are also required to choose at least two classmates' postings and react to their posts. Response postings are required to be between 100 and 150 words and should be thoughtful, substantial, and critique, clarify, and extend the thought being discussed.

As an example, some of the case studies discussed by the students in week 2 included well-known examples of change and configuration management such as the systems integration issues that occurred during the development of the Boeing 787 Dreamliner or the production delays of the Airbus A380. In the latter example, the

TABLE 1.2
Discussion prompts by week

Week	Discussion prompt
1.	No discussion
2.	There are many examples and case studies of improper management of requirements and/or changes that have resulted in catastrophic failures and significant losses for organizations. Find one example that is relevant to change management and share it with the class. Reflect on it and comment in the forum
3.	Properly capturing all application requirements is a key activity for developing the design basis. Application requirements include laws, regulations, and standards, but also social aspects such as cultural differences and behaviors. Find at least two examples of product failures due to differences in culture and share them with the class
4.	In your professional and/or academic experience, do you have structured naming conventions for your files and datasets? If so, briefly describe the naming convention and how it is utilized. Do you (or your organization) use intelligent or non-intelligent ID numbers? What are the strengths and weaknesses?
5.	Many organizations provide customization through web-based tools like product configurators which allow end users to add, change, or tweak certain features and functions of a core product. Find an example of one of these tools and share it with the class. Describe what is unique/different/interesting about the tool and the product itself
6.	Many organizations have personnel dedicated to the processing and management of changes. Find a job announcement related to CM and share it with the class. Describe what you find interesting/exciting/intriguing/challenging about the position
7.	Describe a personal experience you have had, positive or negative, with an organization's help desk. How would you describe the effectiveness or efficiency of the service provided? Was the issue resolved?

digital mock-up used for the design of the electrical routing of the aircraft was not at the required level of robustness for the fuselage. In addition, interoperability issues were identified between design groups using different versions of the CAD software to create the models (CATIA v4 and v5). Other case studies focused on requirements management and miscommunication problems. For example, in 1999, the software controlling NASA's Mars Climate Orbiter's thrusters calculated forces in pounds of force. A separate piece of software, however, took in this data assuming it was in metric units (Newton). Other examples include the recent challenges faced by Spain's rail operator RENFE, where three years after announcing plans to modernize the rolling stock on narrow-gauge commuter trains and medium-distance trains, it was discovered that the trains would be too wide for some of the tunnels. In general, as reported later in their feedback, students agreed that the examination and discussion of these case studies provided a valuable connection to the industrial world where students could observe the link between the material covered in the course and the

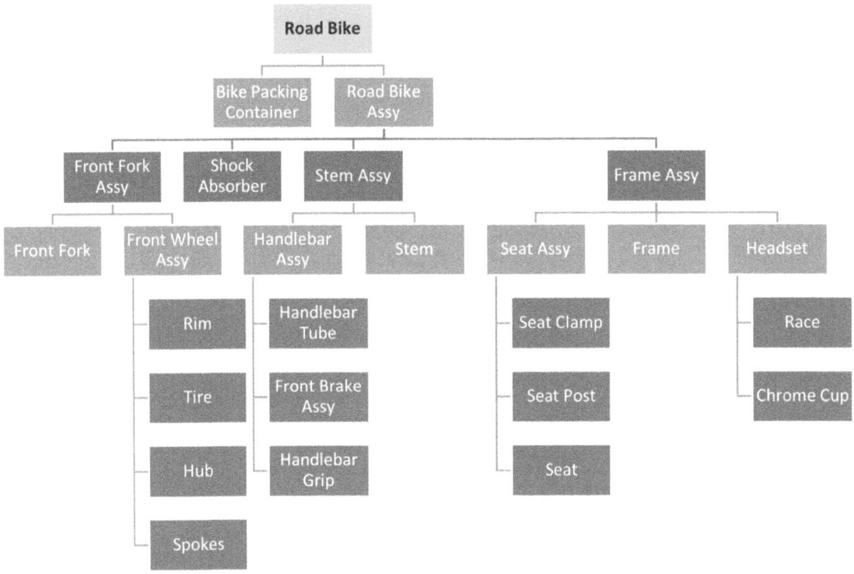

FIGURE 1.3 Product structure of the bicycle used in the assignment.

severely negative consequences that improper change and configuration management practices can have if not considered or executed effectively.

Assignments are practical exercises related to each week's topic. The assignments involve basic calculations and problems using data modeling and Excel in simulated engineering change scenarios. Students create the necessary change objects, calculate the recurring and nonrecurring costs, perform an impact analysis and assessment, determine a change propagates throughout a product hierarchy, and develop a detailed change implementation plan in a "critical path" format, among others. For example, one of the assignments in the course involves the creation of an Investigation Request (IR) and a Change Request (CR). The instructions are as follows: You have entered the business of manufacturing road bicycles. The hierarchical product structure of the bicycle is shown in Figure 1.3.

1. During assembly, your team noticed some challenges while installing the handlebar grip as well as installing the handlebar into the stem. Create an IR form and a corresponding approval workflow for your firm. Populate it with the problem and environment that have been noted.
2. Your bicycle firm has released a prototype bicycle to several riders for feedback. The riders felt that the bicycle was too tall to be comfortable. Using the product structure and assuming a manual change process, populate the simplified CR form shown in Figure 1.4 for decreasing the size of the bicycle. Include all relevant information and the parts that need to be changed. Add rows as needed.

New Change Request			**Date**	
CR Number			**Status**	
CR Description				
Requestor			**Source**	
Proposed Solution				
Cost Estimate			**Priority**	
Tech Review Recommendation				
		Affected Items		
Action	**Old Number**	**Old Revision**	**New Number**	**New Revision**

FIGURE 1.4 Change request form.

COURSE IMPLEMENTATION AND IMPACT

The first iteration of the proposed graduate course was delivered in the fall semester of 2023 and the second iteration in the spring of 2024. A total of ten students in the fall semester and 16 in the spring semester, all working professionals with several years of experience in engineering and technology, enrolled in the course as part of their requirements for their master's degree in engineering technology. This degree program is interdisciplinary and designed for students with an undergraduate degree in engineering, engineering technology, science, information technology, or business.

Because the class is not traditional and all students work full time, a reasonable amount of flexibility was built into the course in terms of submission deadlines. The small class size allowed occasional virtual meetings and office hours with students (one-on-one) to answer questions, clarify certain aspects of an assignment, or provide additional feedback to students, despite some difficulties accommodating schedules and different time zones. Time was also allotted for guest speakers throughout the duration of the course. On several occasions, representatives from the IpX were invited to give presentations and interact with the students. These events served three purposes as follows: first, to further expose our students to change and configuration management practices by listening to industry leaders with firsthand practical experience on the topic; second, to provide networking opportunities for the students interested in potentially pursuing change and configuration management as a career or to learn even more about subsequent CM2 certifications, such as CM2-Professional (CM2-P); and third, to strengthen the existing ties between academia and the industry partner.

The undergraduate version of the course was delivered for the first time in the fall of 2024 with a total enrollment of seven students. Naturally, the content, assignments, and subsequent discussions in this version of the course were adapted to an undergraduate audience by simplifying some of the more complex concepts, reducing the depth of analysis, incorporating more concrete examples, and structuring assignments to build core skills rather than advanced expertise, while still maintaining a challenging academic rigor appropriate for the undergraduate level. In addition, a group project was added to ensure students gained the practical skills necessary to implement and maintain a CM2 standard in a production environment. This approach aligns with other industry standards such as Six Sigma. For example, to be certified as a Green Belt, a student must complete a battery of training coupled with the completion of an actual Six Sigma DMADV (Define, Measure, Analyze, Design, Verify) or DMAIC (Define, Measure, Analyze, Improve, and Control) project within an industry setting, mentored by a Six Sigma Black Belt (Kubiak and Benbow, 2016). Since the audience for CM2 course is comprised of undergraduate students and it would be difficult to secure CM2 implementation projects from industry partners that could be completed in the span of a 16-week semester, a project was devised to allow students to have the necessary level of authority, control, and opportunity for success within the given time constraint.

The project requires involvement with the Purdue Product Data Management (P-PDM) club, a student organization that was recently founded at Purdue University to assist other design and build clubs with the management of their product data by deploying, administering, and supporting all aspects of a PDM system. Many of these engineering design clubs compete at a national level in various events ranging from motorsports, robotics, tractors, autonomous rovers, aviation, and space programs. A common issue with these clubs has historically been their need to effectively manage their digital assets, including requirements, 3D models, simulation data, and bills of materials. Each club has traditionally tackled the management of their data by setting up independent processes and using shared folders and cloud systems such as OneDrive and Google Drive. The transitory nature of these clubs has also posed problems with continuity since graduating seniors in leadership positions within these clubs may or may not leave behind access to the datasets created or stored on personal accounts. Additionally, each club's design processes differ greatly (if they exist at all), and in many cases, the physical build of the intended design often starts well before the datasets have been fully validated. The P-PDM club was established to service these clubs by migrating their datasets to an industrial-level PDM system and improving data integrity and security.

The creation of the P-PDM club was chartered in the fall of the 2023 semester. A group of students established the P-PDM club as a service organization with defined procedures to assist the other clubs similar to how an industry process team would support the various engineering organizations within a company. The club developed a standard service-level agreement (SLA), ticketing processes to intake service requests, formed a process counsel staffed by representatives from each of the clubs supported, and established liaisons between the P-PDM club and the individual clubs they supported to gather requirements and ensure the P-PDM club was meeting the needs of the customer clubs.

The goal of the course project is to assess, plan, and deploy CM2 principles within the P-PDM club so that members have full authority and control of an environment that closely resembles an industry setting. Over the 16-week semester, the students were tasked to take inventory of the P-PDM datasets, create the proper business and product baselines, configure a given PDM system to store the datasets, and implement a common closed-loop change process that will not only be leveraged within the P-PDM club, but also the customer clubs. The project also entails educating other P-PDM club members and customer clubs on CM2. All club datasets are set to be placed under formal change control, including presentations, requirements, supplier information, customer club information, sponsor data, and any other data created in the context of the club.

Each week students are required to provide a project update in the form of a weekly quad chart (see Figure 1.5) to the course instructor who acts as a program manager for the project. Additionally, students are required to give three gate review presentations and report out to other faculty and industry partners on the progress of the project. These two requirements help to simulate the rigor a project of this magnitude would require within an industry setting. As the P-PDM club comes in line with CM2, the club members will then begin servicing customer clubs to perform audits of their datasets and processes and consult with customer club leadership to assist them with implementing formal change control. Some of these activities include identifying change board members and training club members on how to properly release datasets before fabrication. The goal of this undergraduate course going forward is to continually use the P-PDM club as a conduit to progressively bring the P-PDM club and its customer clubs into greater compliance with CM2.

Since the new course covers topics that are typically not part of the traditional engineering and technology curriculum, we were particularly interested in the students' response and their perceptions in terms of the significance and relevance to the industrial world. Since all students have been in the workforce for some time, either as professionals or as interns, we consider their feedback particularly valuable. Student feedback collected at the end of the semester was favorable, which highlights the need and demonstrates the value of this type of skill. Some of the most elaborate responses are included in the next paragraph.

As an example, one student wrote

> I work in the service bill of materials structure composition in my current position in service engineering. I focus on the technical aspects of the serviceability of a unit's components. Ease of access, improvement opportunities, and sometimes reverse engineering to provide feedback to downstream stakeholders (demand planning, field service, knowledge management, etc.). Recently, while working with these process partners, I have learned a lot about the CM2 process at a high level of communication. I will take the official CM2-C certification exam to add to my understanding and knowledge. Overall, this course was exciting because of your in-depth feedback on the assignments. I am a person of appreciation in abundance, so taking this course was an excellent investment in myself and my future!

Similar thoughts were shared and echoed by other students: "This class offered a lot of valuable information that can be applied within my current field of work" and "this class was one of the more impact classes I have taken so far due to how closely it relates with my current line of work and the route our company is trying go down to

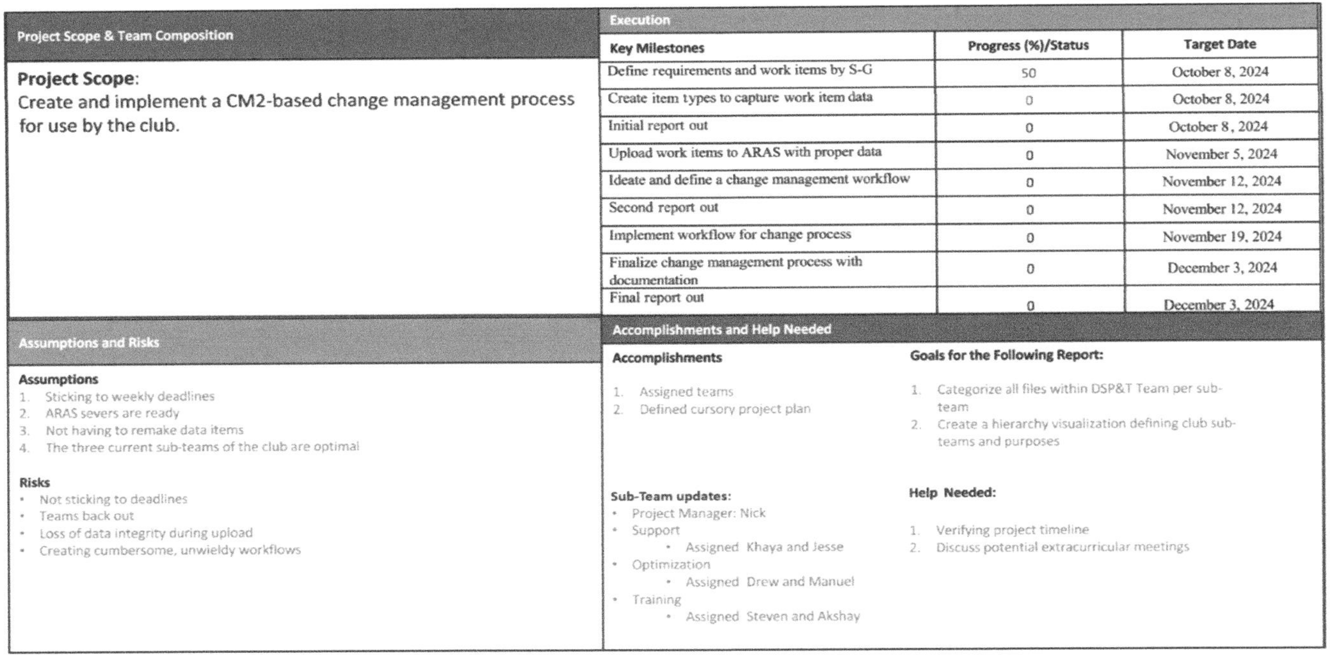

FIGURE 1.5 Weekly quad chart.

improve our processes as an enterprise." Other students wrote: "I was unaware of the importance of configuration management until this class. As time goes on, I will want to know more!" and "This course was a true learning point for me as a professional because of the focus on change management—highlighting its importance. Changes being made in the mass production of deliverables have an impact. This course has helped to detail, describe, and illustrate the types of upgrades, enhancements, and corrections that can be made in those types of productions in the engineering subject matter. Showing purpose in the course readings and course homework/application." Finally, other students reported

> the most impactful thing I learned in this course would be the clear significance of change management in engineering and technology organizations. While I was aware of the change management process, and had participated in change advisory board meetings, I was not fully aware of the full importance of the process. This course has given me the knowledge to encourage others in my organization to embrace the change management process as a crucial part of our ongoing success.

DISCUSSION

The concepts of change and configuration management in general, and CM2 in particular, are a revelation to many who take the professional certificate training courses after having years of experience in an industry setting with context for the benefits that CM2 will provide. Based on the student feedback and comments reported in the previous section, it is safe to say that the proposed course described in this paper had a similar effect on our graduate students. In an undergraduate setting, however, students may only have limited work experience in the form of internships and co-ops, which may not be enough to fully relate to the benefits of proper change and configuration management.

Throughout the implementation of the course, several concerns were identified. The first one is related to the required amount of reading and work that was packed into seven weeks. This concern was voiced by some students who mentioned the fact that in some instances it was difficult for them to balance the required workload for the class while working full time. Since reducing the amount of content is not possible, if we are interested in providing students with the opportunity to become CM2 certified, it is possible that a new version of the course could be developed (either by extending the original course to 8 weeks or making it a regular 16-week course). This format could also provide additional time to work on more examples, activities, and demonstrations and distribute some of the workload across a longer semester. Another concern is the delivery mechanism, which is based primarily on reading assignments, and the lack of examples to illustrate some of the assignments. In this regard, we are considering the creation of short video lectures to support the reading materials as well as posting solutions to sample problems, so students can anticipate the type of responses that are expected in the assignments.

From a content standpoint, several students suggested discussing potential software solutions for managing change (the integration of digital tools and technologies in change management processes) as well as demonstrations of how software could be used to assist in the administration of CM2. Exploring how emerging technologies such as AI, IoT, and data analytics can be leveraged to optimize change management practices was also mentioned in the feedback. In addition, some students pointed out that many of the examples used in the course emphasized mechanical assemblies but did not consider (at least not to the same extent) different products such as chemicals, adhesives, or plastics. Finally, another student suggested that it would have been beneficial to have spent some more time on the roles of the change leader, the change implementation leader, and the change review board. The student wrote

> I feel this section is one of the most important in the course, as it establishes the roles and responsibilities of the team members that will manage change within a business. Missing a stakeholder on a change review board (CRB), for example, could very easily disrupt a company's operations, and many companies, especially large ones, can fall victim to this quite easily if proper process is not followed and/or the proper person with knowledge of internal and external stakeholders is not in charge.

The course described in this paper represents the foundation of a new curriculum focused on digital enterprise education. It covers the theoretical foundations of change and configuration management and the basic elements of the CM2 model. All assignments and activities are performed manually, however. Future work will include the creation of additional courses that will further prepare and equip students with the technical skills and tools necessary to succeed in a digital enterprise environment. For example, we are interested in exposing students to the professional enterprise software systems such as PLM and MRP (Material Requirements Planning) that they will encounter in their future careers and study how change and configuration management processes and workflows are implemented in these environments and how these tools meet the requirements of change management standards. Other additions will include guest lectures from industry professionals.

CONCLUSION

Many authors agree that establishing and maintaining effective, sustainable, and mutually beneficial relationships between academic and industry professionals are key elements for shaping a competitive workforce. Indeed, the industrial world cannot simply rely on academia to produce the perfect workforce without an active collaboration (Wohlin et al., 2011; Peters and Lucietto, 2016). Academia-industry collaboration may take many different forms, such as internships, advisory boards, involvement in capstone courses, and research partnerships. Although the proactive involvement and initiative of the academic collaborators are fundamental, it is equally important for the industrial partner to allocate the necessary resources and commitment to foster and nurture the relationship. In this regard, communication and shared responsibility are key. From a pedagogical standpoint, no curriculum or teaching method will fully prepare students for the real-world leap without a clear understanding of industry needs.

As engineering and manufacturing companies continue their transformation toward becoming digital enterprises, finding skilled professionals who understand this new way of working can be challenging. The demand for data scientists, AI specialists, and engineers who can fully leverage digital technology will far exceed the supply. In this regard, companies need to invest in training and talent development.

The course discussed in this paper is the result of a productive industry-academia partnership and collaboration. The IpX works with global academic institutions to support industrial education and research partnerships and offers students holding CM2 certification job placement, awareness, and community. Through this collaborative effort, university students gained access to the CM2 methodology, the most robust framework and quality management platform for creating and maintaining a digital enterprise with complete end-to-end ecosystem traceability. CM2 training teaches students practical and actionable hands-on methods preparing them for the workforce and providing credentials to their resumes to set themselves apart in the hiring process.

REFERENCES

Agost, M.J. Company, P. Contero, M., and Camba, J.D. (2022). CAD training for digital product quality: A formative approach with computer-based adaptable resources for self-assessment. *International Journal of Technology and Design Education*, 32(2), 1393–411. https://doi.org/10.1007/s10798-020-09651-5

Aranburu, A., Cotillas, J., Justel, D., Contero, M., and Camba, J. D. (2022a). How does the modeling strategy influence design optimization and the automatic generation of parametric geometry variations? *Computer-Aided Design*, 151, 103364. https://doi.org/10.1016/j.cad.2022.103364

Aranburu, A. Justel, D. Contero, M., and Camba, J.D. (2022b). Geometric variability in parametric 3D models: Implications for engineering design. *Procedia CIRP*, 109, 383–388. https://doi.org/10.1016/j.procir.2022.05.266

Arthur, L. J. (1988). *Software Evolution: The Software Maintenance Challenge*. Wiley-Interscience.

Bartlett, K.A., and Dorribo Camba, J. (2023). The role of a graphical interpretation factor in the assessment of spatial visualization: A critical analysis. *Spatial Cognition & Computation*, 23(1), 1–30. https://doi.org/10.1080/13875868.2021.2019260

Boyd, L. Lu, Y., and Souppez, J.B. (2023). Pedagogy 4.0: Employability skills and computer aided design (CAD) education for Industry 4.0. 28th International Conference on Automation and Computing, Birmingham, United Kingdom, 2023, 01–06. https://doi.org/10.1109/ICAC57885.2023.10275222

Camba, J.D., Contero, M., and Hartman, N. (2021). The cost of change in parametric modeling: A roadmap. *Computer-Aided Design and Applications*, 18, 634–643. https://doi.org/10.14733/cadaps.2021.634-643

Camba, J.D. Fuerst, T., Ewing, D., and Anderson, J. (2023). *The CM2 Approach to Change and Configuration Management*. eAcademicBooks. ISBN: 978-0-9962452-9-6.

Camba, J., and Anderson, J. (2024). Change and configuration management as a foundation for a new digital enterprise education. In: Mrugalska, B., Karwowski, W., and Trzcielinski, S. (eds) Human Aspects of Advanced Manufacturing, Production Management and Process Control. AHFE (2024) International Conference, 146. AHFE International, USA. http://doi.org/10.54941/ahfe1005147

Chechile, I. (2021). Complexity Creep. *NewSpace Systems Engineering*, 119–135. Springer. https://doi.org/10.1007/978-3-030-66898-3_4

CIMdata (2019). Need For a 21st Century Engineering Curriculum (Commentary). www.cimdata.com/en/resources/complimentary-reports-research/commentaries/item/11528-need-for-a-21st-century-engineering-curriculum-commentary.

Dart, Susan A. (1992). The Past, Present, and Future of Configuration Management. Carnegie Mellon University. https://doi.org/10.1184/R1/6585185.v1

Eckert, C., and Clarkson, J. (2023). The Evolution of Complex Engineering Systems. In: *Handbook of Engineering Systems Design*, 1–39. Springer International Publishing. https://doi.org/10.1007/978-3-030-46054-9_6-2

Estublier, J., Leblang, D., Hoek, A.V.D.., Conradi, R., Clemm, G., Tichy, W., and Wiborg-Weber, D. (2005). Impact of software engineering research on the practice of software configuration management. *ACM Transactions on Software Engineering and Methodology*, 14(4), 383–430. https://doi.org/10.1145/1101815.1101817

Farayola, O.A., Hassan, A.O., Adaramodu, O.R., Fakeyede, O.G., and Oladeinde, M. (2023). Configuration management in the modern era: best practices, innovations, and challenges. *Computer Science & IT Research Journal*, 4(2), 140–157. https://doi.org/10.51594/csitrj.v4i2.613

Frame, R. (2004). Integrating PLM methods into the undergraduate curriculum. 2004 Annual Conference & Exposition, Salt Lake City, UT, 9–764. https://doi.org/10.18260/1-2--13329

Fuerst, T., Camba, J.D., and Mazumdar, A. (2024). Engineering graphics education for the digital enterprise: A practical example in a large freshman engineering course. 2024 ASEE Annual Conference & Exposition, Portland, OR. https://doi.org/10.18260/1-2--47283

Gromova, E.A. (2020). PLM as a sequential round of the technological revolution. IOP Conference Series: Materials Science and Engineering, 968(1), 012027. https://doi.org/10.1088/1757-899X/968/1/012027

Hamraz, B. Caldwell, N.H., and Clarkson, P.J. (2013). A holistic categorization framework for literature on engineering change management. *Systems Engineering*, 16(4), 473–505. https://doi.org/10.1002/sys.21244

IEEE (2012). IEEE Standard for Configuration Management in Systems and Software Engineering, in IEEE Std 828–2012 (Revision of IEEE Std 828-2005), 1–71, 16 March 2012.

Institute for Process Excellence – IpX (2024). The CM2 Platform. https://ipxhq.com/training/cm2-certification-courses/the-CM2-platform.

ISO 10007:2017 – Quality Management – Guidelines for Configuration Management. www.iso.org/standard/70400.html

Jarratt, T.A. Eckert, C.M. Caldwell, N.H., and Clarkson, P.J. (2011). Engineering change: an overview and perspective on the literature. *Research in Engineering Design,* 22, 103–124. https://doi.org/10.1007/s00163-010-0097-y

Johnson, N. Reidy, L. Droll, M., and LeMon, R.E. (2012). Program Requirements for Associate's and Bachelor's Degrees: A National Survey. Complete College America.

Jokic, M.D., Wandel, A.P., and Kist, A.A. (2010). Which version art thou? Configuration management in engineering education. In Proceedings of the 21st Annual Conference of the Australasian Association for Engineering Education (AAEE 2010). Sydney, Australia, Dec. 5–8, 2010. https://research.usq.edu.au/item/q01z6/which-version-art-thou-configuration-management-in-engineering-education

Jones, M.D. Hutcheson, S., and Camba, J.D. (2021). Past, present, and future barriers to digital transformation in manufacturing: A review. *Journal of Manufacturing Systems*, 60, 936–948. https://doi.org/10.1016/j.jmsy.2021.03.006

Kubiak, T.M., and Benbow, D.W. (2016). *The Certified Six Sigma Black Belt Handbook.* Quality Press.

Lindkvist, C., Stasis, A., and Whyte, J. (2013). Configuration management in complex engineering projects. *Procedia CIRP*, 11, 173–176. https://doi.org/10.1016/j.procir.2013.07.046

Morris, J., and Wagner, J.R. (2023). Application of extracurricular course teaching product lifecycle management concepts to undergraduates . ASEE Annual Conference & Exposition, Baltimore, MD. https://doi.org/10.18260/1-2--42269

Nasir, M.H.N., and Sahibuddin, S. (2011). Critical success factors for software projects: A comparative study. *Scientific Research and Essays*, 6(10), 2174–2186.

Nyemba, W.R. Mbohwa, C., and Carter, K.F. (2021). Systems thinking research: adapting for engineering change management. In: *Bridging the Academia Industry Divide: Innovation and Industrialisation Perspective using Systems Thinking Research in Sub-Saharan Africa*, 41–56. Springer International Publishing. https://doi.org/10.1007/978-3-030-70493-3_3

Oz, E., and Sosik, J.J. (2000). Why information systems projects are abandoned: a leadership and communication theory and exploratory study. *Journal of Computer Information Systems*, 41(1), 66–78. https://doi.org/10.1080/08874417.2000.11646978

Peters, D.L., and Lucietto, A.M. (2016). A survey of types of industry-academia collaboration. 2016 ASEE Annual Conference & Exposition, New Orleans, LA. https://doi.org/10.18260/p.26455

Philbeck, T., and Davis, N. (2018). The Fourth Industrial Revolution: Shaping a new era. *Journal of International Affairs*, 72(1), 17–22.

SAE International (2019). Configuration Management Standard EIA649C. www.sae.org/standards/content/eia649c

Sandrin, E., Forza, C., Leitner, G., and Trentin, A. (2022). Configuration manager: describing an emerging professional figure. In Proceedings of the 26th ACM International Systems and Software Product Line Conference-Volume B, 193–200. https://doi.org/10.1145/3503229.3547049

Schmidt, R., Lyytinen, K., Keil, M., and Cule, P. (2001). Identifying software project risks: An international Delphi study. *Journal of Management Information Systems*, 17(4), 5–36. https://doi.org/10.1080/07421222.2001.11045662

United States Department of Defense (USDoD, 1992). Military Standard Configuration Management (MIL-STD-973). https://apps.dtic.mil/sti/citations/ADA297556.

Whyte, J., Stasis, A., and Lindkvist, C. (2016). Managing change in the delivery of complex projects: Configuration management, asset information and 'big data.' *International Journal of Project Management*, 34(2), 339–351. https://doi.org/10.1016/j.ijproman.2015.02.006

Wohlin, C., Aurum, A., Angelis, L., et al. (2011). The success factors powering industry-academia collaboration. *IEEE Software*, 29(2), 67–73. https://doi.org/10.1109/MS.2011.92

2 Analysis of the Evolution of Human Factors in Industry 4.0 to 5.0

From Security and Automation to Advanced Collaboration and Well-being

Mariannys Rodríguez Gasca, Sanderson Cesar Macedo Barbalho, and Eugenia Cornils Monteiro da Silva

INTRODUCTION

As technologies advance, industrial environments undergo significant transformations in their organizational structures, processes, and consequently, in the role of the human factor (Neumann et al., 2021). The advent of Industry 4.0 (I4.0), defined by its collaborative nature that promotes interaction between humans and machines, has been a pivotal driver of this change. Although I4.0 technologies were initially presented as an umbrella concept encompassing intelligent automation, data exchange, and smart manufacturing systems (Gladysz et al., 2023), early efforts largely concentrated on the development and implementation of these technologies. This narrow focus raised concerns about the potential marginalization of the human factor (Adam et al., 2019; Ligarski et al., 2021).

In this context, attention was focused almost exclusively on the development and application of technology, which raised concerns about the possible displacement of the human factor. However, the emergence of evidence on the limitations of a technology-only approach has prompted a shift in perspective toward a more human approach that considers the needs and capabilities of people in advanced technological environments. For example, studies such as that of (Macias-Aguayo et al., 2022) explored the barriers and facilitators of I4.0 and Lean Six Sigma integration, highlighting how high costs and insufficient training were initial barriers that prevented the effective adoption of technologies, ultimately leading to a re-evaluation of the role of the human factor. Similarly, Vuksanović Herceg et al. (2020) highlighted

DOI: 10.1201/9781003625360-2

that, during digital transformations, organizations often perceive human resources as hurdles rather than catalysts, particularly when employees lack the necessary skills and competencies.

Moreover, other studies (Cascio & Montealegre, 2016; Fletcher et al., 2020; Kong, 2019) underscore how neglecting the human dimension in the transition to I4.0 can lead to adverse outcomes, such as unfavorable working conditions, inadequate safety standards, and diminished employee well-being. These findings collectively underscore the importance of addressing human factors to achieve a balanced and sustainable integration of I4.0 technologies.

These challenges emphasized the importance of employee training, skills, and competencies to ensure effective and sustainable adoption of emerging technologies (Kadir & Broberg, 2020; Rodriguez Gasca & Macedo Barbalho, 2024). Organizations began to recognize that human knowledge, creativity, and experience were irreplaceable (Casalino et al., 2021). This shift in vision recognizes that workers are not only operators, but also decision-makers, problem-solvers, and key contributors in an increasingly dynamic and complex industrial environment (Thamm et al., 2021). Rana and Rathore (2023) mention that the adoption of I4.0 also highlighted the creation of a new educational paradigm that focuses on the development of the intelligent workforce, intelligent operators, intelligent materials, and intelligent project managers; this approach promoted the concept of Education 4.0. Training and coaching have become capable moderators to mitigate the adverse effects of the implementation of technologies and enhance their competitive advantages (Galanti et al., 2023).

There are several studies that address the factors associated with the management and development of the human factor in this context. Mattsson et al. (2020) point out that in the future, operators will need to be prepared to manage a variety of work situations and deal with complex interactions. Al Amri et al. (2021) mention that academic and professional qualifications turn out to be key factors in the adaptation process to the new I4.0 methodologies. Kowal et al. (2022) conducted a study where they identified the main digital, technical, social, and personal skills of employees that are required in the context of I4.0. The authors also highlight, as a main result, that I4.0 is highly technical, and technical and soft skills are equally important. From this same perspective, authors such as Marmier et al. (2021) propose some learning tools to measure the degree of maturity of companies and develop the skills required for employee training.

In contrast, systems incorporating I4.0 technologies not only require advanced technical and cognitive skills from users, but also offer powerful tools to facilitate continuous learning, professional training, and support in performing complex tasks, both from a physical and cognitive point of view (Merkel et al., 2019). Technologies such as augmented reality (AR) and virtual reality (VR) have proven effective in training workers in simulated environments that mimic real-life conditions, allowing for risk-free practice and enhancing their skills (Asai et al., 2012; Heine et al., 2023). Collaborative robots, known as cobots, work alongside human operators, taking on repetitive or particularly demanding tasks, to help reduce the complexity of the workload (Colim et al., 2021; Fournier et al., 2024). On the other hand, we also find smart sensors, wearables, and other connected devices that collect and analyze data in real time, allowing flexibility and scalability while providing critical information that improves decision-making and efficiency (Merhar et al., 2019; Roda-Sanchez et al., 2019). Finally, more recently, technologies associated with artificial intelligence, such as industrial chatbots, virtual assistants, deep learning algorithms, and mainly ChatGPT, have significantly transformed the knowledge economy, being adopted in various sectors, to support learning and decision-making processes (Alier et al., 2025; Jauhar et al., 2024).

In light of these advances, the aim of this paper was to present a literature review on the human factor in the adoption of I4.0 technologies, covering studies published from 2011 to 2023. A systematic literature search approach was used to identify relevant publications and ensure the inclusion of significant and current studies. In addition, the documents collected in the systematic search were subjected to a bibliometric analysis. Specialized tools such as VOSviewer and some histograms provided by the Scopus platform were used to generate maps of co-authorship networks, keyword co-occurrence, citation analysis, and publication behavior during this time period, in order to visualize emerging patterns or gaps in existing research.

METHODOLOGY

The literature review conducted in this study was executed in two conceptual steps: a systematic search and a bibliometric analysis of the articles. This analysis adhered to several recommended guidelines to ensure a high-quality review, including bibliographies characterized by recent publication dates, high citation frequencies, and publications with significant impact factors (Uemura Reche et al., 2020).

Systematic Search Protocol

The search process was conducted using a trial-and-error approach, as suggested by (Booth et al., 2016; Hussain et al., 2023), where relevant keywords were identified to achieve the desired results. Following this, combinations of keywords that yielded the most effective outcomes in the search were determined. Thus, after several attempts, it was possible to establish (TITLE-ABS-KEY ("Human Factor") AND TITLE-ABS-KEY ("Industry 4.0" OR digitalization)) as the combination that addressed documents with the greatest contribution to the subject. The term "Human Factor" was used only because it is a more specialized term that could show more focused results in studies that analyze the relationship between people and technological systems, industrial processes, or work environments, which is particularly relevant for research in areas such as I4.0 and human–machine interaction.

The next step was the definition of the Scopus database since it was the one that best adapted to the needs of the research once it provided access to a wide range of documents from different scientific journals. The Scopus filter of title, abstract, and keywords was applied for the first search. In addition to these filters, the Boolean operators AND and OR were used to specify that both terms should appear in the content of the sample documents. This first stage resulted in a total of 587 documents. As we wanted to identify the behavior of research relating to the terms human factor and I4.0 from its beginnings to the present, the search was limited to documents from 2011, when the term I4.0 was first introduced in Germany (Xu et al., 2021), until the end of 2023. Additionally, these results were refined by type of document, limiting the search to article, review, and conference paper. As a result, a total of 448 documents were obtained.

Bibliometric Analysis Protocol

Bibliometrics deals with the quantitative analysis of publications, articles, and citations, providing statistically relevant data on the production and dissemination of knowledge in various areas. This approach allows for the objective assessment of scientific and academic activity (Donthu et al., 2021).

In this work, bibliometrics is used to extract information from the selected articles and determine an overview of the behavior of research according to the time period, countries, authors, areas of study, citations, and occurrence of keywords, capable of contributing to a bibliometric study. In this context, the study is divided into two main sections focused on bibliometric analysis: (1) analysis of authors and behavior of publications and (2) analysis of key terms.

For the analysis of keywords, an analysis was performed using the bibliometric software VOSviewer. Such analysis helps both to analyze the content of the publications and to investigate patterns and trends in the study area. The results of this phase of keyword network maps were used to identify more important research topics in the study area. In addition, the behavior of the occurrences of keywords was analyzed throughout the analyzed period of time.

RESULTS

This section presents and discusses the main findings related to the proposed methodology and extracted from the sample of 448 documents. The results of the selected documents provide quantitative and qualitative information on the behavior and trends of the studied topic that can serve as a contribution to the state of the art.

Behavior of the Number of Documents per Year

Figure 2.1 shows how the number of documents published on the topic has behaved from 2011 to May 2023. It is possible to notice a variation in the number of articles produced in each period, with an increasing trend over the years. In particular, the cumulative percentage line shows that 80% of the research on the topics studied was published in the last four years. In contrast, the years prior to 2017 have very few publications, probably because this concept was still being fully introduced and developed. A significant relative increase in the percentage of publications was also observed in 2019, compared to 2018. In this period, the total number of publications went from 19 documents to 52 documents, in other words, an increase of 8%, which suggests a turning point in research on the human factor and I4.0.

In general terms, these results could be a reflection of technological advances, changes in the industry, an increase in funding for research on this topic, or even external factors such as the emergence of the pandemic, as discussed and

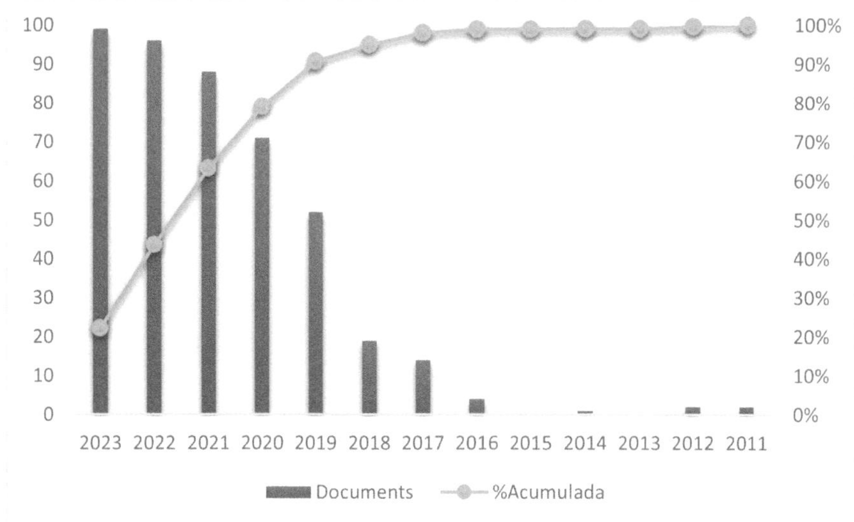

FIGURE 2.1 Evolution and cumulative distribution of annual publications on human factors and I4.0. (Source: Author.)

corroborated by the high degree of digitalization suggested in (CEPAL, 2020). On the other hand, the fact that the highest percentage of publications is concentrated in recent years also suggests a great potential for innovation in the way in which I4.0 technologies are designed to integrate with human factors within industrial environments.

Behavior of the Number of Publications by Country/Territory

Regarding contributions by country/territory, Figure 2.2 shows the 12 countries with the most documents, out of a total of 68 countries, which have contributed at least 13 articles out of the total of 448 publications analyzed on the topic. These countries represent a contribution of 80% of the research in the total sample studied. The figure shows a strong concentration of publications in a few countries, suggesting that research on the topic is predominantly driven by certain research centers in Europe (Italy, Germany, and the Russian Federation). This could indicate that international collaboration still has room to expand. In particular, it is observed that, of the three main countries in the figure, Italy and the Russian Federation are predominantly producing research by national institutions, while Germany is shown as one of the countries with a higher tendency to collaborate with other countries. The United States, the United Kingdom, and Canada also stand out as countries with a significant amount of collaboration.

On the other hand, the analysis of these countries in terms of the number of citations confirms that Italy and Germany are among the most influential countries in the subject; however, the Russian Federation does not stand out much in this regard.

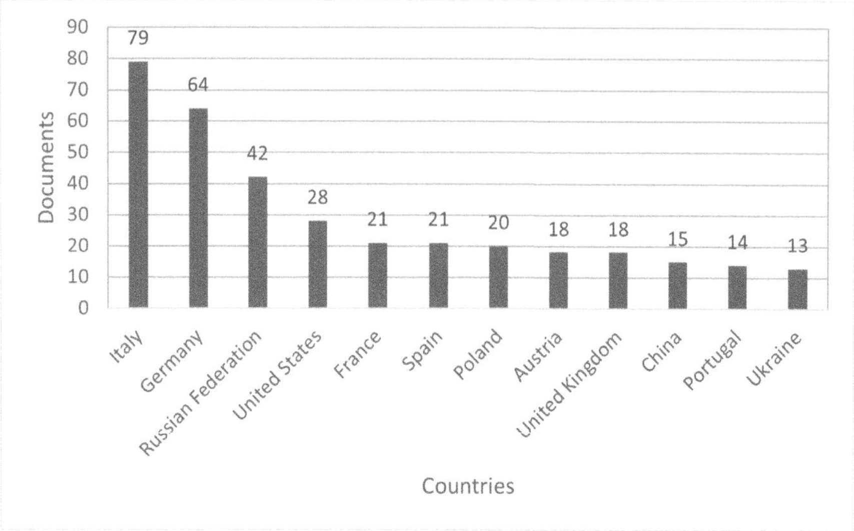

FIGURE 2.2 Documents by country/territory. (Source: Author.)

In contrast, countries such as Canada and France add up to a significant number of citations of their documents. In particular, Brazil appears lower on the list, in position 19, with a total of 8 documents. Although the number of documents is not as robust, it has a significant number of citations. In addition, Brazil is positioned as the South American country that marks an important performance in the area.

Additionally, Figure 2.3 presents a network visualization that represents international research collaborations, with each node representing a country and the connections between them indicating co-authorship or joint research projects. It is likely that the visualization software used has selected this specific time period to focus on the years in which there is relevant or significant data according to the metric of documents per country. Countries such as Germany, Italy, and the United States have a higher degree of collaboration activity, as they are more central and connected to multiple other nodes. The network reveals a strong concentration of collaboration among European countries, reflecting the well-established research networks within the European Union. The United States also appears as a major hub, demonstrating its global reach in research partnerships.

Emerging connections can be observed between countries outside the traditional centers of research, such as Brazil and India, which appear connected but less central. Russia and its neighboring countries, like Ukraine and Hungary, show more isolated subnetworks, indicating regional collaboration patterns. We see those countries, such as Germany and Italy, highlighted as relevant in this field, had their peak of publications in the period of 2020, while countries such as India, Indonesia, and the Netherlands have had their most significant contributions in the period of 2022. These results suggest to us that the topic has indeed evolved and continues to be increasingly explored in other countries.

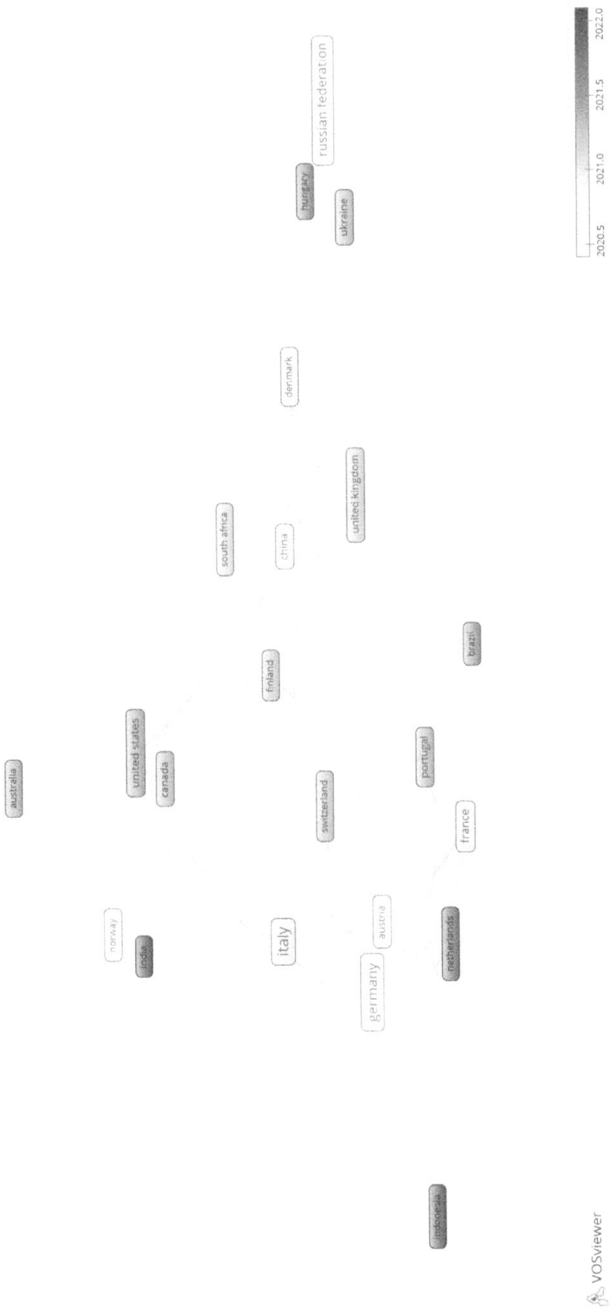

FIGURE 2.3 Graph visualizing the correlation of countries by documents published over time. (Source: Author.)

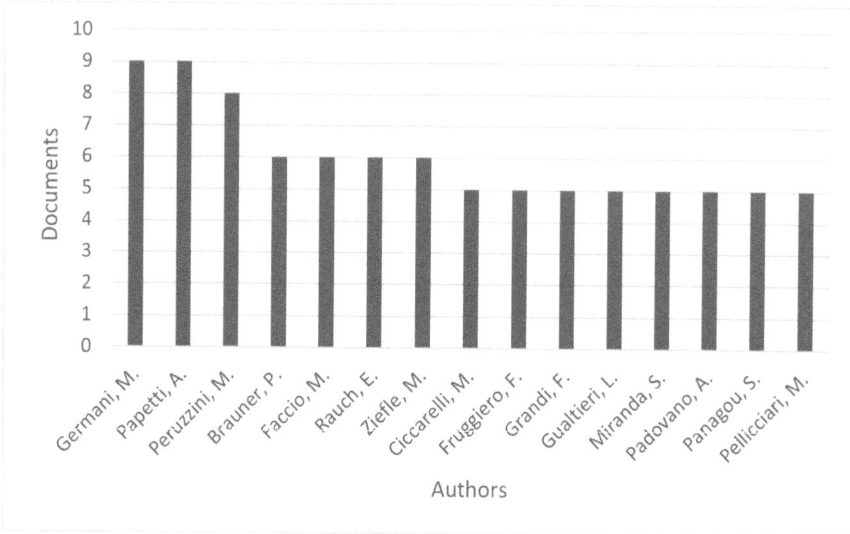

FIGURE 2.4 Distribution graph of the most relevant authors according to the number of publications. (Source: Author.)

The difference in contributions could suggest, beyond investment in research, differences between important regional dynamics such as research support policies and management of the interaction between academia and industry for studies dedicated to I4.0. Consequently, countries with more contributions could be an indicator that they better manage the development of these fields and the level of maturity of the concept of I4.0 in their territories.

Analysis of Relevant Authors

For the author analysis, the number of contributions of each author in the total number of documents in the sample was evaluated. Figure 2.4 shows a bar chart that organizes the top 15 authors with the most contributions in documents published in the area. This group of authors contributes approximately 20% of the total documents analyzed on the topic. Those leading the group are "Germani, M.," "Papetti, A.," and "Peruzzini, M.," with a total of eight to nine publications each in the last 13 years. This could indicate that these researchers are leaders in the field, with a solid track record of publications and possibly with relevant collaborations and projects. Authors with a similar number of publications are usually involved in collaborative projects. It should also be noted that the entire group of these authors belongs to Italian institutions.

Additionally, in order to correlate the importance of authors by document number with the impact of the publications, a compilation of the ten articles with the highest number of citations out of the total of 448 documents in the sample is presented in Table 2.1. These documents can be considered as having a high impact factor on the topic associated with the human factor and I4.0; therefore, the inclusion of these articles

TABLE 2.1
Ten relevant authors by impact of the publications

Authors	Title	Cites	Journal/Source	Year
Francesco Longo, Antonio Padovano, and Steven Umbrello	Value-oriented and ethical technology engineering in Industry 5.0: A human-centric perspective for the design of the factory of the future	327	*Applied Sciences (Switzerland)*	2020
W. Patrick Neumann, Sven Winkelhaus, Eric H. Grosse, and Christoph H. Glock	I4.0 and the human factor – A systems framework and analysis methodology for successful development	317	*International Journal of Production Economics*	2021
Luca Gualtieri, Erwin Rauch, and Renato Vidoni	Emerging research fields in safety and ergonomics in industrial collaborative robotics: A systematic literature review	247	*Robotics and Computer-Integrated Manufacturing*	2021
Marie-Pierre Pacaux-Lemoine, Damien Trentesaux, Gabriel Zambrano Rey, and Patrick Millot	Designing intelligent manufacturing systems through human-machine cooperation principles: A human-centered approach	239	*Computers and Industrial Engineering*	2017
Bzhwen A. Kadir, Ole Broberg, and Carolina Souza da Conceição	Current research and future perspectives on human factors and ergonomics in Industry 4.0	201	*Computers and Industrial Engineering*	2019
Lara Bartocci Liboni, Luciana Oranges Cezarino, Charbel José Chiappetta Jabbour, Bruno Garcia Oliveira, and Nelson Oliveira	Smart industry and the pathways to HRM 4.0: implications for SCM	180	*Supply Chain Management*	2019
Fabio Sgarbossa, Eric H. Grosse, W. Patrick Neumann, Daria Battini, and Christoph H. Glock	Human factors in production and logistics systems of the future	180	*Annual Reviews in Control*	2020
Luca Silvestri, Antonio Forcina, Vito Introna, Annalisa Santolamazza, and Vittorio Cesarotti	Maintenance transformation through Industry 4.0 technologies: A systematic literature review	174	*Computers in Industry*	2020

(continued)

TABLE 2.1 (Continued)
Ten relevant authors by impact of the publications

Authors	Title	Cites	Journal/Source	Year
Margherita Peruzzini, Fabio Grandi, and Marcello Pellicciari	Exploring the potential of Operator 4.0 interface and monitoring	158	*Computers and Industrial Engineering*	2020
Sandra Grabowska, Sebastian Saniuk, and Bożena Gajdzik	Industry 5.0: Improving humanization and sustainability of Industry 4.0	143	*Scientometrics*	2022

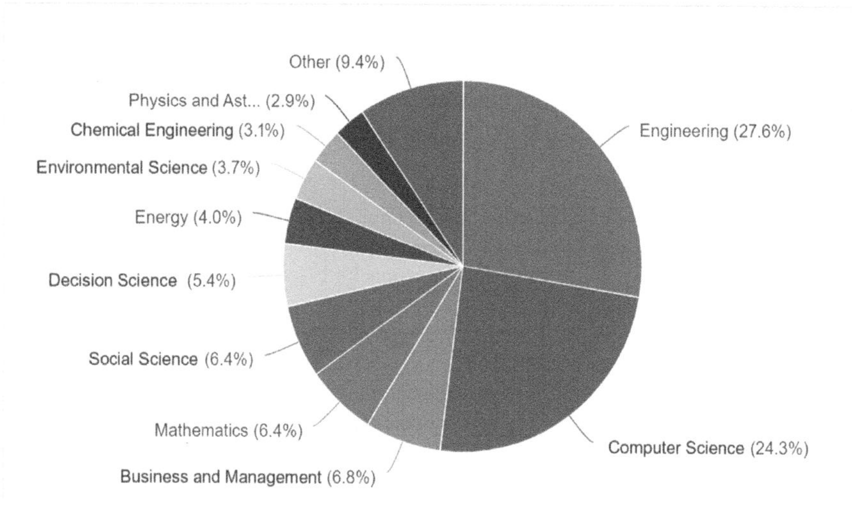

FIGURE 2.5 Document distribution graph by area of study. (Source: Author.)

in the references of a literature review on the topic is considered important, because they would form the basis of scientific research on the topic in the period 2011–2023.

It was confirmed that several of the authors with a high number of publications have documents with a considerable number of citations. This may indicate a deep dedication and specialization in the area and high-quality and relevant research valued by the academic community and that could be key to describing the evolution of the topic and future development paths.

Document Production by Area of Study

Figure 2.5 presents, using a pie chart, the percentage of documents by field of study. Engineering and computer science areas dominate research in this field, with a share of 27.6% and 24.3%, respectively, of the total publications. This demonstrates the crucial role that these disciplines play in the creation of advanced technologies, such as intelligent automation systems and data analysis tools. These contributions enable more efficient and safer production, while transforming interactions between workers and machines. However, this technical approach poses challenges, especially in the adaptation of workers to complex technological environments, where greater attention is required to human aspects such as well-being, ergonomics, and skills development.

These challenges are mitigated by the interdisciplinary nature of research, as it was observed that collaborations between areas such as engineering and computer science with other disciplines are quite common. This synergy is necessary to enrich the integration of advanced technologies in the workplace, addressing not only the technical aspects, but also the challenges related to the interaction between workers and automated systems. Despite this, areas such as medicine and psychology represent only a small percentage of 9.4% of the publications. This is crucial, as it could indicate a possible gap in research from the perspective of human behavior and the mental and physical health of workers.

Keyword Cluster Analysis

For the keyword co-occurrence analysis, only terms that occurred at least 10 times were included. We identified 3377 terms in the titles, abstracts, and keywords, of which 68 met the threshold. To understand the graph, it is necessary to know that the node size is proportional to the number of occurrences of keywords, while the links represent the index of occurrence. The software calculates the total strength of the co-occurring links with other keywords for each term and selects the words with the highest total link strength. In addition, the marked divisions in the co-occurrence network represent thematic clusters. Terms within the same cluster are thematically related and are grouped based on the frequency with which they appear together, so they could suggest some trend or line of research.

Figure 2.6 shows the result of the keyword network of the sample studied in this work and the five clusters associated with the sample. These results show some trends that are important to highlight. Starting with the first cluster, it shows that there is a frequent association between some specific technologies of I4.0, such as AR, VR, and big data, with terms such as engineering education, personnel training, ergonomics, safety engineering, risk management, productivity, product design, and maintenance. This could indicate that technologies such as AR, VR, and big data are seen not only as technological tools but also as key resources to train workers and prepare future generations of engineers for the environment enabled by I4.0. In turn, training the human factor through these technologies seems to have the purpose of improving risk management at work and increasing productivity. It can also be observed that applications in maintenance activities seem to be the most explored in this regard. Finally, the association of some of the terms of this cluster, such as product design and ergonomics, suggests studies focused on the end user and the user experience.

Another interesting observation can be made based on the analysis of the second cluster. We found that there is a correlation of occurrences between the terms human resource management (HRM), man–machine system, human collaborative robot, safety, accident prevention, risk assessment, collaborative robots, assembly,

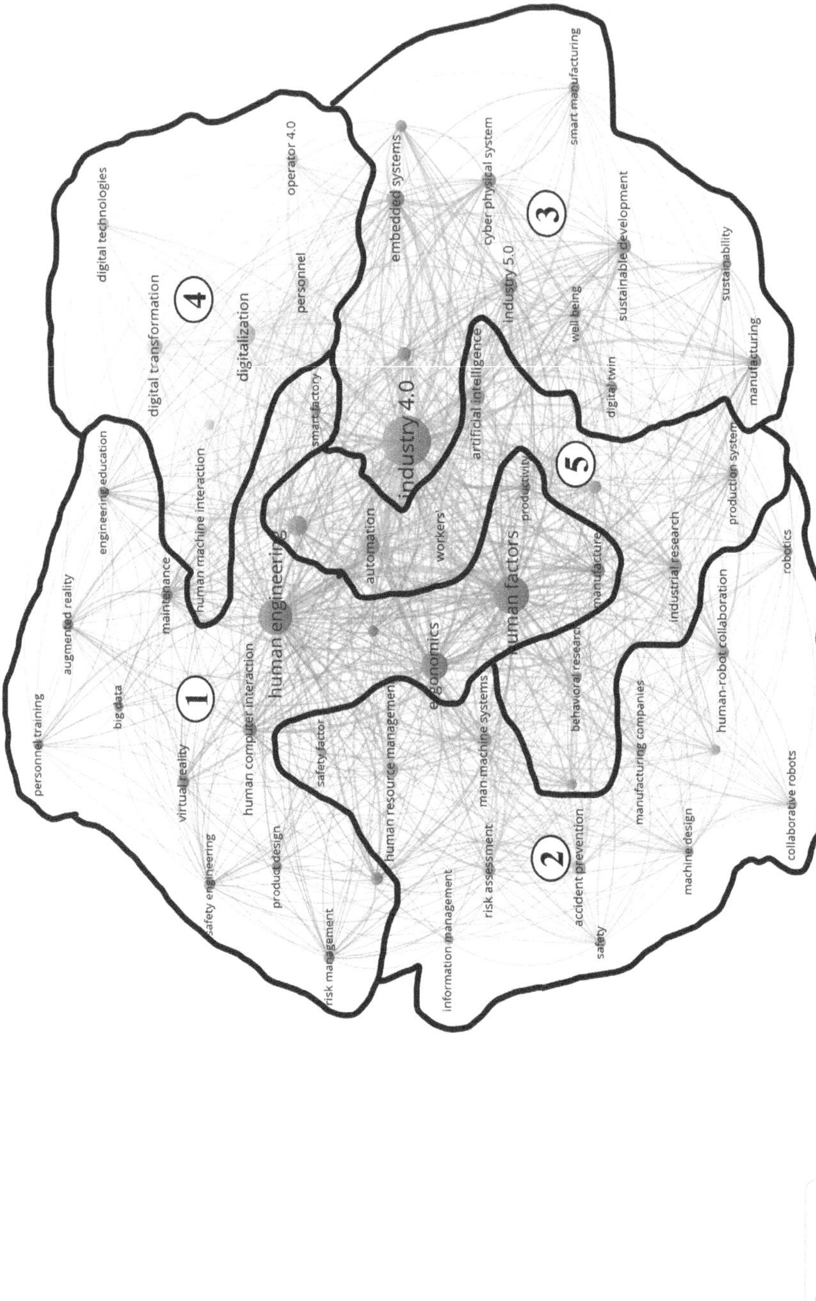

FIGURE 2.6 Network graph of keyword co-occurrence clusters. (Source: Author.)

manufacturing companies, information management, and machine design. This cluster suggests research focused on managing the safe integration of collaborative robotics in industry, where robot–human collaboration in assembly processes is the main focus, as it can increase efficiency and reduce the risk of errors and accidents. In this cluster, machine design could be a key element to facilitate effective collaboration in these systems; however, this term suggests a more technical approach.

Regarding the third cluster, it shows an association between technologies such as digital twins, embedded systems, cyber–physical systems, and the Internet of things, with terms such as well-being, Industry 5.0, and sustainable development. From the association of these terms, it could be said that research studies the integration of the aforementioned technologies for responsible production that seeks to balance economic efficiency with environmental and social impact and well-being. In addition, the association with the term Industry 5.0 suggests a movement focused on the reintegration of the human element in industrial automation, but from a more autonomous point of view that allows the capabilities of the human factor to be enhanced instead of replacing them.

In the fourth cluster, the terms that appear together frequently are Operator 4.0, workers, efficiency, digital technologies, and human–machine interaction. This cluster of terms suggests a transformation in the role of workers within I4.0 and is described by the term Operator 4.0, which indicates that training in new technologies and adaptability are essential for job performance. Finally, the fifth cluster shows a correlation between the terms decision-making, cognitive systems, decision support systems, production systems, automation, and artificial intelligence. This integration indicates research focused on the move toward smarter factories where decision-making is strongly supported by advanced technologies such as cognitive systems integrated with artificial intelligence.

Analysis of Keyword Occurrences by Time Period

Figure 2.7 presents a graph of keyword occurrences during the time period analyzed in this study. The bibliometric visualization software automatically shows a specific time period in the bottom bar, in this case, 2020 to 2021, as it is the most significant according to the occurrence metric; that is, most occurrences are concentrated in this period. The light colors of the nodes indicate terms whose highest concentration of occurrences are older, while dark colors are terms whose occurrences are more recent. These results allow us to visualize the evolution of research interest in the area of human factor and I4.0.

As important points, analyzing the initial phases of the graph, approximately between 2019 and 2020, terms such as automation, human–computer interaction, embedded systems, product design, risk management, big data, and staff training dominate the occurrences. During this period, the focus was on process optimization and the integration of digital technology in product management and design. The priority was to improve operational efficiency and manage risks associated with the adoption of new technologies. As time moves forward to more recent periods, around 2021 and 2022, an occurrence of terms such as digital twin, artificial intelligence,

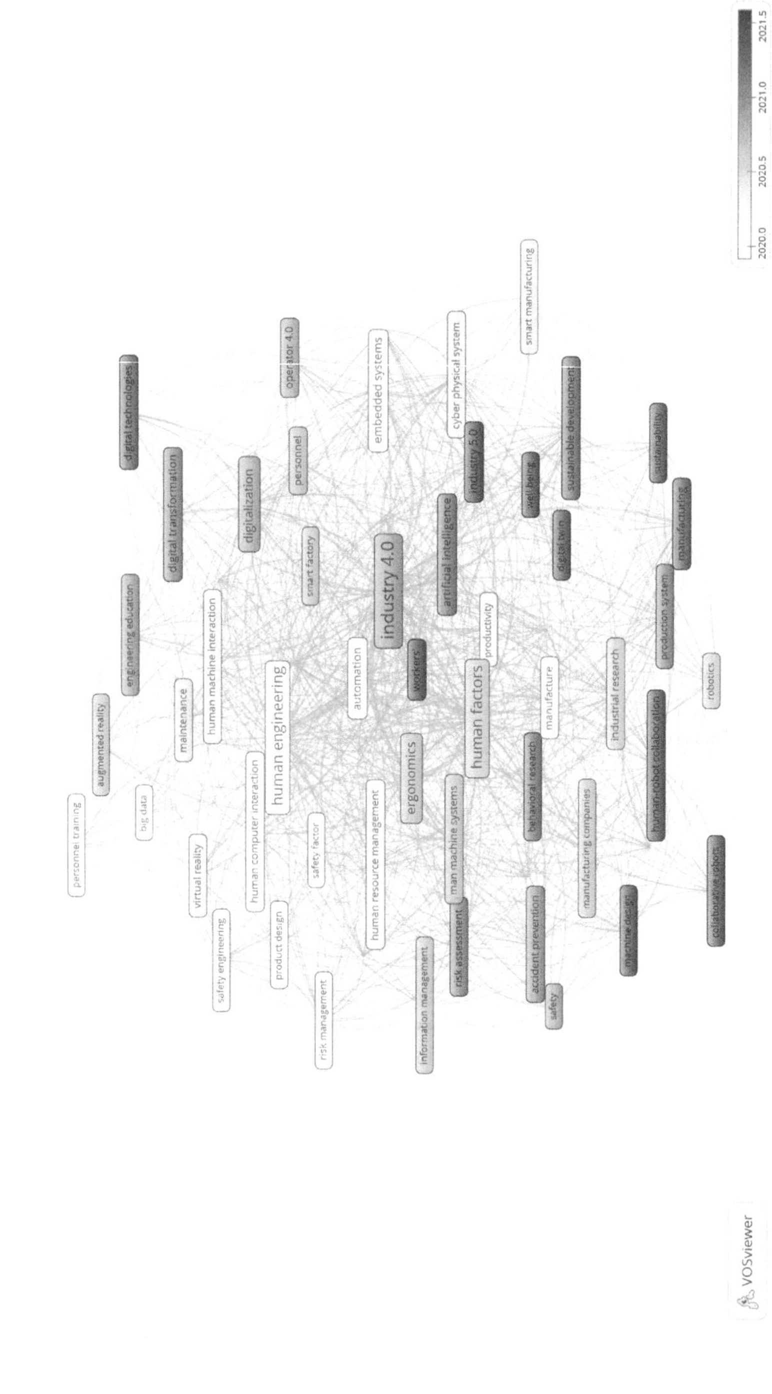

FIGURE 2.7 Time period graph with the highest concentration of keyword occurrences.

manufacturing, human–robot collaboration, machine design, Industry 5.0, workers, and sustainability is observed. In this period, the emerging themes reflect an evolution in the focus of research and practice from basic automation and technological integration toward the adoption of more advanced technologies and innovative concepts. The introduction of "digital twin" and "artificial intelligence" indicates a focus on simulation and machine learning technologies, while "Industry 5.0" and "sustainability" suggest a shift toward integrating human capabilities with advanced technologies and greater awareness of environmental impact.

Lines of Research Over Time

Table 2.2 shows the main research topics and trends of the analyzed documents according to the clusters of occurrences of keywords and the occurrences by time period. For the analysis, the results were divided into two parts. The research lines were observed in the time period between 2011 and 2018, and then, the research lines were observed in the period from 2019 to 2023. It was divided in this way because a significant increase in the number of documents was observed starting in 2019.

The results allow us to observe the behavior of how research trends on the topic have evolved over the last 14 years. Starting with publications from 2011 to 2018, it was observed that research focuses mainly on the transition of both production systems and the human factor toward automated environments, where cyber–physical systems, but with a strong focus on human–machine interaction, are a dominant trend. Studies are also addressed on the challenges of HRM, development of new skills, and training strategies to ensure that workers adapt and collaborate with advanced technologies. Particularly, this trend is widely explored in 2018, where it is also observed

TABLE 2.2
Evolution of the main topics on the human factor and I4.0

Timespan	Main topics
2011–2018	• Automation, cyber–physical systems, and robots • Man–machine integration • Education for I4.0 • Ergonomics for safety • Worker performance • Operator 4.0
2019–2023	• Human–robot collaboration • Cognitive ergonomics • Personalization and well-being • Risk assessment and management • Increased use of AR, VR, and digital twin • Assistance systems and cognitive systems • Automation of decision-making • Industry 5.0 and Operator 5.0 • Sustainability • Job satisfaction

that the term Operator 4.0 appears more strongly and there is greater concern for issues associated with occupational safety.

On the other hand, in the period from 2019 to 2023, a significant focus is observed on human-centered design, with an emphasis on the customization of work environments, integrating advanced technology such as AR and VR, and implementation of wearable devices. Studies suggest that in addition to concerns about the physical health of workers, mental health also becomes an important focus in high-tech environments and has gained relevance, with studies focusing on risk assessments and measures to mitigate the negative impact on cognitive load and improve well-being in digitalized factories. This in turn is associated with the emergence of research on applications of assistance systems, support systems, and cognitive systems. Likewise, in this period, the automation of decision-making has emerged as a key trend, highlighting the integration of systems based on artificial intelligence and machine learning. These studies are associated with the analysis of large volumes of data in real time, to improve accuracy by providing recommendations based on emerging patterns and trends that could go unnoticed by humans. On the other hand, around 2020 and 2021, the term Industry 5.0 emerged, strongly associated with the idea of maximizing both the benefits of I4.0 technologies and the characteristics of the human factor. This term is also strongly associated with the focus of studies on social sustainability, probably driven by the impact of COVID-19, as the world had to rethink the structures of existing working methods. This same movement could also have influenced the conduct of research on the adaptation of the human factor to technologies.

CONCLUSION

The findings suggest that topics associated with the human factor and I4.0 have been increasing since the term was introduced in 2011 until 2023, with a significant increase starting in 2019, which could have been driven by the impact of the changes brought about by COVID-19. On the other hand, European countries dominate the topic with a greater number of documents and more influential authors in the area. However, the results suggest that the outlook for Brazil seems to be improving, especially in the last five years.

The results also suggest some major differences in the topics of interest of the research over the time period analyzed. Firstly, in the period between 2011 and 2018, the focus of the studies was on integrating humans into automated systems, ensuring the safety and reliability of interactions. While in the period from 2019 to 2023, the focus evolved toward elements of more advanced collaboration between humans and robots. Ergonomics went from focusing solely on improving safety and work performance to a more comprehensive approach to worker health, also concerned with emotional well-being, cognitive load, and personalization of the work environment. From the second period mentioned, technologies such as AR, VR, digital twin, and artificial intelligence have been consolidated as tools in applications ranging from training, assistance in complex tasks, decision support, and cognitive systems. Research also moved from focusing solely on automation to worrying about how workers accept and interact with these technologies, a key factor for the success of digital transformation. Throughout the first period of time, the human factor was present, but more as a component of the interaction with machines, while with the emergence of the concept of Industry 5.0, it is reintroduced into the center of the industrial process. This marks a significant evolution from a purely technical and productive vision to a more humanized and responsible vision of the use of technology.

This research allowed us to analyze the behavior of studies on human factors and I4.0 technologies in production systems. However, there are some methodological limitations, since the systematic search was limited to a single database, which was Scopus. Future work could include a larger database and perform a more detailed content analysis of the documents.

ACKNOWLEDGEMENTS

The authors would like to acknowledge the contribution of the Federal District Research Support Foundation (FAP-DF), as well as the University of Brasilia and the Postgraduate Program in Mechatronic Systems (PPMEC), for their financial support.

REFERENCES

Adam, C., Aringer-Walch, C., & Bengler, K. (2019). *Digitalization in Manufacturing – Employees, Do You Want to Work There?* (pp. 267–275). https://doi.org/10.1007/978-3-319-96068-5_30

Al Amri, T., Puskas Khetani, K., & Marey-Perez, M. (2021). Towards Sustainable I4.0: Key Skill Areas for Project Managers in GCC Construction Industry. *Sustainability*, *13*(15), 8121. https://doi.org/10.3390/su13158121

Alier, M., Pereira, J., García-Peñalvo, F. J., Casañ, M. J., & Cabré, J. (2025). LAMB: An open-source software framework to create artificial intelligence assistants deployed and integrated into learning management systems. *Computer Standards & Interfaces*, *92*, 103940. https://doi.org/10.1016/j.csi.2024.103940

Asai, S., Ogawa, T., & Takebayashi, H. (2012). Visualization and Digitization of Welder Skill for Education and Training. *Welding in the World*, *56*(9–10), 26–34. https://doi.org/10.1007/BF03321378

Booth, A., Sutton, A., & Papapioannou, D. (2016). *Systematic Approaches to a Successful Literature Review* (second edition). Sage.

Casalino, A., Zanchettin, A. M., Piroddi, L., & Rocco, P. (2021). Optimal Scheduling of Human–Robot Collaborative Assembly Operations with Time Petri Nets. *IEEE Transactions on Automation Science and Engineering*, *18*(1), 70–84. https://doi.org/10.1109/TASE.2019.2932150

Cascio, W. F., & Montealegre, R. (2016). How Technology Is Changing Work and Organizations. *Annual Review of Organizational Psychology and Organizational Behavior*, *3*(1), 349–375. https://doi.org/10.1146/annurev-orgpsych-041015-062352

CEPAL. (2020). *Informe Especial COVID-19 No 4: las empresas frente a la COVID-19: emergencia y reactivación*. Comisión Económica para América Latina y el Caribe (CEPAL). www.cepal.org/es/publicaciones/45734-sectores-empresas-frente-al-covid-19-emergencia-reactivacion

Colim, A., Morgado, R., Dinis-Carvalho, J., & Sousa, N. (2021). An Empirical Study of the Work Conditions and Productive Performance after Collaborative Robotics Implementation in a Manufacturing Assembly Process. *FME Transactions*, *49*(4), 859–866. https://doi.org/10.5937/fme2104859C

Donthu, N., Kumar, S., Mukherjee, D., Pandey, N., & Lim, W. M. (2021). How to Conduct a Bibliometric Analysis: An Overview and Guidelines. *Journal of Business Research*, *133*, 285–296. https://doi.org/10.1016/j.jbusres.2021.04.070

Fletcher, S. R., Johnson, T., Adlon, T., Larreina, J., Casla, P., Parigot, L., Alfaro, P. J., & Otero, M. del M. (2020). Adaptive Automation Assembly: Identifying System Requirements for Technical Efficiency and Worker Satisfaction. *Computers & Industrial Engineering*, *139*, 105772. https://doi.org/10.1016/j.cie.2019.03.036

Fournier, É., Jeoffrion, C., Hmedan, B., Pellier, D., Fiorino, H., & Landry, A. (2024). Human-Cobot Collaboration's Impact on Success, Time Completion, Errors, Workload, Gestures and Acceptability During an Assembly Task. *Applied Ergonomics*, *119*, 104306. https://doi.org/10.1016/j.apergo.2024.104306

Galanti, T., De Vincenzi, C., Buonomo, I., & Benevene, P. (2023). Digital Transformation: Inevitable Change or Sizable Opportunity? The Strategic Role of HR Management in Industry 4.0. *Administrative Sciences*, *13*(2), 30. https://doi.org/10.3390/admsci13020030

Gladysz, B., Tran, T., Romero, D., van Erp, T., Abonyi, J., & Ruppert, T. (2023). Current Development on the Operator 4.0 and Transition Towards the Operator 5.0: A Systematic Literature Review in Light of Industry 5.0. *Journal of Manufacturing Systems*, *70*, 160–185. https://doi.org/10.1016/j.jmsy.2023.07.008

Heine, I., Hellebrandt, T., Huebser, L., & Padrón, M. (2023). Hybrid Intelligence: Augmenting Employees' Decision-Making with AI-Based Applications. In *Handbook of Human-Machine Systems* (pp. 321–332). Wiley. https://doi.org/10.1002/9781119863663.ch27

Hussain, M., Zhang, T., & Seema, M. (2023). Adoption of Big Data Analytics for Energy Pipeline Condition Assessment – A Systematic Review. *International Journal of Pressure Vessels and Piping*, *206*, 105061. https://doi.org/10.1016/j.ijpvp.2023.105061

Jauhar, S. K., Sethi, S., Kamble, S. S., Mathew, S., & Belhadi, A. (2024). Artificial Intelligence and Machine Learning-Based Decision Support System for Forecasting Electric Vehicles' Power Requirement. *Technological Forecasting and Social Change*, *204*, 123396. https://doi.org/10.1016/j.techfore.2024.123396

Kadir, B. A., & Broberg, O. (2020). Human Well-Being and System Performance in the Transition to Industry 4.0. *International Journal of Industrial Ergonomics*, *76*, 102936. https://doi.org/10.1016/j.ergon.2020.102936

Kong, F. (2019). Development of Metric Method and Framework Model of Integrated Complexity Evaluations of Production Process for Ergonomics Workstations. *International Journal of Production Research*, *57*(8), 2429–2445. https://doi.org/10.1080/00207543.2018.1519266

Kowal, B., Włodarz, D., Brzychczy, E., & Klepka, A. (2022). Analysis of Employees' Competencies in the Context of Industry 4.0. *Energies*, *15*(19), 7142. https://doi.org/10.3390/en15197142

Ligarski, M. J., Rożałowska, B., & Kalinowski, K. (2021). A Study of the Human Factor in Industry 4.0 Based on the Automotive Industry. *Energies*, *14*(20), 6833. https://doi.org/10.3390/en14206833

Macias-Aguayo, J., Garcia-Castro, L., Barcia, K. F., McFarlane, D., & Abad-Moran, J. (2022). Industry 4.0 and Lean Six Sigma Integration: A Systematic Review of Barriers and Enablers. *Applied Sciences*, *12*(22), 11321. https://doi.org/10.3390/app122211321

Marmier, F., Deniaud, I., Rasovska, I., & Michalak, J.-L. (2021). Towards a Proactive Vision of the Training for the 4.0 Industry: From the Required Skills Diagnostic to the Training of Employees. *IFAC-PapersOnLine*, *54*(1), 1144–1149. https://doi.org/10.1016/j.ifacol.2021.08.135

Mattsson, S., Fast-Berglund, Å., Li, D., & Thorvald, P. (2020). Forming a Cognitive Automation Strategy for Operator 4.0 in Complex Assembly. *Computers & Industrial Engineering*, *139*, 105360. https://doi.org/10.1016/j.cie.2018.08.011

Merhar, L., Berger, C., Braunreuther, S., & Reinhart, G. (2019). *Digitization of Manufacturing Companies: Employee Acceptance Towards Mobile and Wearable Devices* (pp. 187–197). https://doi.org/10.1007/978-3-319-94619-1_18

Merkel, L., Berger, C., Braunreuther, S., & Reinhart, G. (2019). *Determination of Cognitive Assistance Functions for Manual Assembly Systems* (pp. 198–207). https://doi.org/10.1007/978-3-319-94619-1_19

Neumann, W. P., Winkelhaus, S., Grosse, E. H., & Glock, C. H. (2021). Industry 4.0 and the Human Factor – A Systems Framework and Analysis Methodology for Successful Development. *International Journal of Production Economics*, *233*, 107992. https://doi.org/10.1016/j.ijpe.2020.107992

Rana, B., & Rathore, S. S. (2023). Industry 4.0 – Applications, Challenges and Opportunities in Industries and Academia: A Review. *Materials Today: Proceedings*, *79*, 389–394. https://doi.org/10.1016/j.matpr.2022.12.162

Roda-Sanchez, L., Olivares, T., Garrido-Hidalgo, C., & Fernández-Caballero, A. (2019). *Gesture Control Wearables for Human-Machine Interaction in Industry 4.0* (pp. 99–108). https://doi.org/10.1007/978-3-030-19651-6_10

Rodriguez Gasca, M., & Macedo Barbalho, S. (2024). *Exploring the Influence of Industry 4.0 Technologies on Workplace Dynamics: A Literature Analysis*. https://doi.org/10.54941/ahfe1005149

Thamm, S., Huebser, L., Adam, T., Hellebrandt, T., Heine, I., Barbalho, S., Velho, S. K., Becker, M., Bagnato, V. S., & Schmitt, R. H. (2021). Concept for an Augmented Intelligence-Based Quality Assurance of Assembly Tasks in Global Value Networks. *Procedia CIRP*, *97*, 423–428. https://doi.org/10.1016/j.procir.2020.05.262

Uemura Reche, A. Y., Canciglieri Junior, O., Estorilio, C. C. A., & Rudek, M. (2020). Integrated Product Development Process and Green Supply Chain Management: Contributions, Limitations and Applications. *Journal of Cleaner Production*, *249*, 119429. https://doi.org/10.1016/j.jclepro.2019.119429

Vuksanović Herceg, I., Kuč, V., Mijušković, V. M., & Herceg, T. (2020). Challenges and Driving Forces for Industry 4.0 Implementation. *Sustainability*, *12*(10), 4208. https://doi.org/10.3390/su12104208

Xu, X., Lu, Y., Vogel-Heuser, B., & Wang, L. (2021). Industry 4.0 and Industry 5.0—Inception, Conception and Perception. *Journal of Manufacturing Systems*, *61*, 530–535. https://doi.org/10.1016/j.jmsy.2021.10.006

3 Control and Task Scheduling Algorithms for Automated Guided Vehicles in Industry 4.0

Marcin Mrugalski

INTRODUCTION

The rapid development of autonomous vehicles supported by the development of artificial intelligence algorithms enabling the processing of a significant amount of data from sensors, improvements in areas of electric drives, advancements in wireless communication, and control systems has led to the widespread use of automated guided vehicles (AGVs) in many areas of the economy (Sun et al., 2022; Vlachos et al., 2023; Witczak et al., 2022; Zhang et al., 2023b). Over the past decade, there have been significant changes in the industry caused by, among others, the use of transport systems based on AGVs in industry, which in many cases made it possible to move away from the traditional production process based on the production line in favor of flexible production systems and agile manufacturing. Despite these advancements, the deployment of AGVs presents numerous challenges. Commonly encountered issues include optimal route selection or route planning (Zaharia et al., 2020). Moreover, issues related to traffic planning, in particular obstacle detection, collision avoidance, resolving deadlocks, and congestion, are important problems (Ming et al., 2024). Another problem is issues related to AGV localization and management. Moreover, effective battery management and addressing vehicle failures are vital research domains (Lipiec et al., 2021; Witczak et al., 2020a). Therefore, novel approaches are needed to maintain the performance of transportation systems, commonly comprising multiple AGVs. Among various challenges, AGV scheduling and control demand specific focus. In deterministic systems, where influencing factors like transport and loading/unloading times are fixed, an offline control system is applicable. Nonetheless, these factors are frequently unpredictable or inherently stochastic in practice. Thus, real-time, online decision-making control systems are crucial. In order to solve such a significant problem, a centralized model predictive control (MPC) algorithm for AGV fleet scheduling is proposed in this work.

DOI: 10.1201/9781003625360-3

THE PROBLEM OF TASK SCHEDULING

Task scheduling problems are optimization tasks focused on determining a global order for task execution, considering time restrictions, operational priorities, and inter-task dependencies within the process. The most common areas of application of task scheduling algorithms include production processes (Vlachos et. al, 2023), warehouse systems (Zhang et al., 2023b), and transport processes (Sun et al., 2022). The expected result of task scheduling is the optimal task schedule with respect to the adopted criterion. This criterion may vary depending on the task, and the most common criteria include minimizing energy consumption, the number of vehicles used (Zou et al., 2022), minimizing delay, minimizing makespan, maximum flowtime, total resources used, and the sum of tardiness for AGV (Riazi et al., 2020). Multicriteria criteria are often used. Scheduling can be defined as a decision-making procedure involving the distribution of resources to tasks across specific time intervals, aiming to optimize certain objectives. Resources, tasks, and objectives can vary widely; for example, objectives may include minimizing the completion time of the final task or reducing overall delay, based on the context. Problems in the area of scheduling concern concepts such as tasks and resources. A task is defined as performing a sequence of actions, which are called operations. Operations, on the other hand, require the implementation of specific resources. The resources include technical devices, materials or raw materials, and electricity used to perform the task. However, in the case of tasks, one can mention the readiness date, the expected completion date of the task, the possibility of interrupting or dividing the operation, and possible ways of performing the activity. Resources are often divided into renewable, nonrenewable, and doubly limited. Scheduling algorithms should also be characterized by certain desirable properties, such as divisibility, task tolerance, scalability, flexibility, and responsiveness.

Task scheduling methods can be divided into deterministic and stochastic approaches. Deterministic scheduling implies that all parameter values are precisely known in advance, irrespective of their availability time. Consequently, these values remain constant throughout system execution. Stochastic scheduling involves parameters whose exact values might not be predetermined. Nevertheless, these values are not entirely unknown and are typically represented using fuzzy sets or stochastic functions. Moreover, depending on the availability of parameter values, all scheduling problems can be classified into offline or online ones. In offline scheduling, all parameters and their values are predetermined prior to system execution. Whereas, in online scheduling, complete parameters are not initially available, and they emerge during execution. For instance, jobs are introduced over time without prior knowledge of their release timing or specific attributes. Furthermore, the scheduling procedures can ultimately be categorized by the methods used to construct schedules as static or dynamic. Static scheduling involves creating a schedule before system execution. This method necessitates the availability of all job parameters beforehand; otherwise, the schedule cannot be executed as initially designed. Dynamic scheduling involves creating the schedule progressively as the system operates. These methods typically make only the next necessary decision without fully developing the schedule beforehand. Dynamic scheduling is frequently employed in flexible manufacturing systems

(FMSs) (Mrugalski, 2023; Zacharia and Xidias, 2020) due to shared resource usage, route flexibility, and the stochastic arrival of raw materials. In the last decade, much research has been carried out to solve the problem of assigning tasks to a fleet of AGVs. This is a vast area due to the huge variety of tasks and types of constraints involved. Table 3.1 presents an example of the division of task scheduling methods for AGVs.

TABLE 3.1
Methods for AGV scheduling

Class of scheduling methods	Scheduling methods	Articles
Exact methods	Mixed integer linear programming (MILP)	(Fatemi-Anaraki et al., 2023;Fayazi and Vahidi, 2018)
	Branch and bound (B&B)	(Espinouse et al., 2000; Wang and Zeng, 2022)
	Exhaustive search approach (ESA)	(Gil-Gala et al., 2020)
	Constraint programming (CP)	(Bocewicz et al., 2009; Novas and Henning, 2014)
	Model predictive control (MPC)	(Majdzik et al., 2023)
Artificial intelligence	Reinforcement learning (RL)	(Hu et al., 2020; Zhang et al., 2024)
	Artificial neural network (ANN)	(Wang et al., 2022b; Zhang, 2023a)
Heuristic	Dispatching rule (DR)	(Nageswara et al., 2017; Vivaldini et al., 2015)
	Job shop scheduling problem (JSSP)	(Adams et al., 1988; Yuan, 2023)
	Parallel two-step decomposition-based heuristic (PTSDBH)	(Amirteimoori et al., 2023)
Metaheuristic	Simulated annealing (SA)	(Chen et al., 2019; Xiang et al., 2023)
	Tabu search (TS)	(Seo et al., 2007; Zheng et al., 2014)
	A*	(Santos et al., 2021)
	Variable neighborhood search (VNS)	(Sze et al., 2016)
	Evolutionary algorithms (EA)	(Luo et al., 2023)
	Bio-inspired intelligence (BI)	(Chen et al., 2022)
Game theory	Multidimensional evolutionary games (MEG)	(Xu et al., 2024)
Approximation methods	Polynomial time approximation scheme (PTAS)	(Li et al., 2022)
	Fully polynomial time approximation scheme (FPTAS)	(Jeong, 2024)

3.3 TASK SCHEDULING FOR AGVS IN TRANSPORT SYSTEMS IN INDUSTRY 4.0

Transport systems based on AGV are used in Industry 4.0, especially in FMSs. In this case, AGVs deliver the components necessary for production to the appropriate assembly stations and move the finished products to the receiving stations, from where they are then transported to the appropriate locations in the warehouse. In such a situation, the sequence of tasks performed and the timely transport of components/products are important, because another component cannot be delivered to the place, for example, of assembly, if the previously manufactured product has not been taken from that place. In this case, the aspect of waiting for the product to be collected and the issue of controlling the movement of transport vehicles are important. Figure 3.1 shows stages/subtasks which may occur during the implementation of a transport system.

It should be emphasized that the problem of task scheduling is one of the most complex and critical for the operation of the entire transport system. As part of task scheduling, the optimal allocation of the transport task to a specific AGV among all available vehicles should be made.

The solution to the problem of scheduling transport tasks is a pair of vectors $\{x,t\}$ where $x=(x_1,\ldots,x_n)$ is a vector of transport tasks, while $t=(t_1,\ldots,t_n)$ is a vector of time moments in which the implementation of individual tasks begins. It should be emphasized that a solution that satisfies all the constraints in the scheduling problem

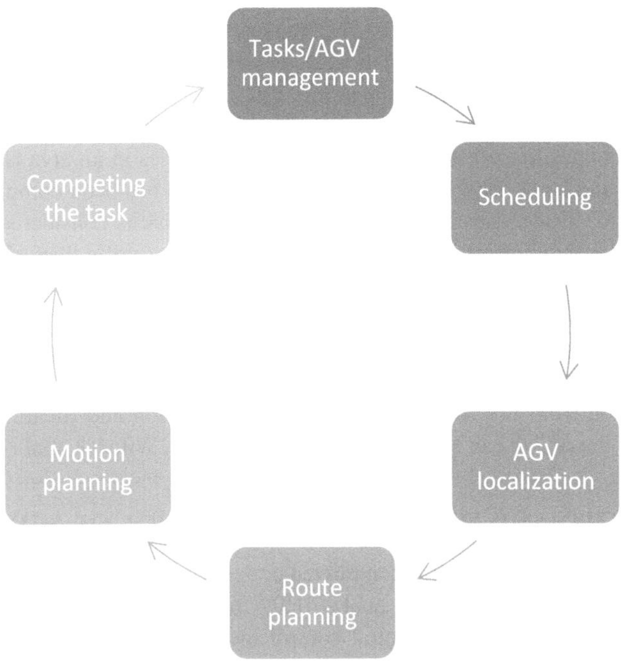

FIGURE 3.1 Stages in the transport system.

is called an admissible solution. The optimization task consists of determining an admissible solution for which the assumed objective function reaches the optimal value where the objective function represents the assumed quality of the system's functioning (Smutnicki et al., 2022). In other words, the task of optimization in the transport system is to determine the best solution in terms of the adopted criterion, taking into account existing constraints. The most frequently used task scheduling algorithms are presented in the previous chapter. At the localization stage, a resource is identified and located in the work area, for example, AGV, which enables the implementation of a given task. At this stage, the strategy of selecting the nearest vehicle enabling the transport task is usually used. However, when allocating a vehicle, existing constraints should be taken into account, that is, time constraints, priority constraints, and mobility constraints (Shi et al., 2020). The presence of constraints means that in practice the allocation of transport tasks in AGV fleet systems can become quite complex. Nevertheless, the literature contains a number of interesting methods/strategies for selecting vehicles to perform a given task, taking into account, for example, the state of charge of vehicle batteries or the remaining useful life of the vehicle. These types of strategies allow us to reduce the costs of operating the transport system. Once the task is assigned to the AGV resource, path planning takes place, that is, the process of determining the optimal transport path to the destination. During path planning, the other AGVs should be located and potential obstacles detected in order to avoid collisions. Path planning is based on a prepared representation of the environment in which AGVs move to search for sequences of operations in order to reach the destination point in an optimal way, for example, in terms of time, distance, or minimum amount of energy. When the path planning algorithm determines the optimal route, it is possible to start the transport task. During its implementation, motion planning may occur, consisting of real-time modification of a planned path according to dynamic obstacles appearing. The motion planning mechanism seeks to prevent deadlock scenarios and minimize vehicle density in specific areas to prevent traffic congestion. Restricting the number of vehicles in an area is called zonal control (Ho and Liao, 2009). Once a collision-free path has been completed and the vehicle has reached its destination, the transportation task is considered finished. In addition to the abovementioned problems related to the use of a fleet of AGVs, there is the problem of vehicle management (Branisso et al., 2012). It monitors the current technical condition of vehicles and the degree of their degradation. In this phase, the remaining useful life of the AGV is determined before it is sent to the service center. At this stage, faults or failures of AGVs may also occur, which may lead to their exclusion from use. In the literature, Fault Detection and Identification (FDI) methods for the AGVs can be found (Lipiec et al., 2021; Witczak et al., 2020b).

CENTRALIZED AND DECENTRALIZED CONTROL OF AGV TASKS

Task scheduling can be done in a centralized or decentralized manner using an appropriate control algorithm (De Ryck et al., 2020; Chen et al., 2022). A centralized control system is one in which a central control mechanism has control directly over lower-level components of the system or indirectly through the use of a hierarchical management structure. This type of control is characterized by complex behavior

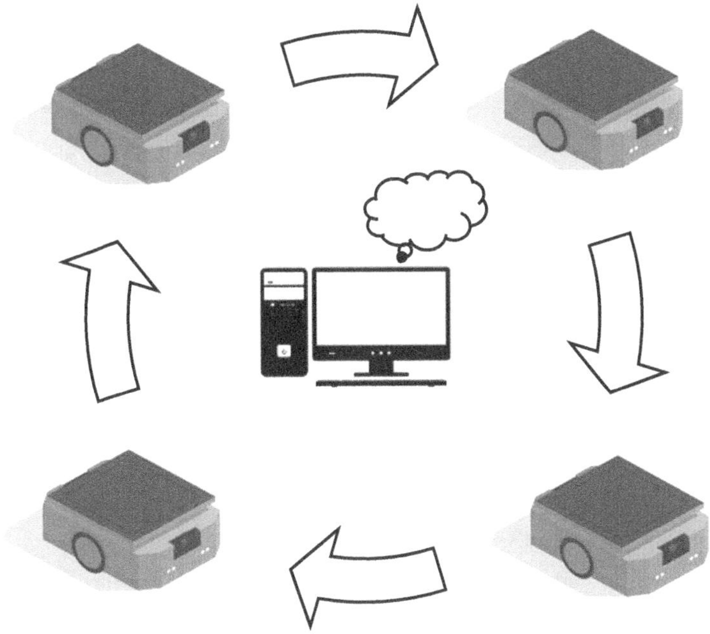

FIGURE 3.2 Centralized control.

resulting from the need to supervise lower-level elements. The literature contains numerous methods enabling centralized control of a fleet of AGVs (De Ryck et al., 2020). Figure 3.2 shows the centralized control scheme for AGVs.

A decentralized control system is a system in which lower-level components make decisions independently using local information in order to achieve the assumed global goals (c.f. Figure 3.3). The literature contains numerous methods enabling decentralized control of a fleet of AGVs (De Ryck et al., 2020; Draganjac et al., 2016).

Both control methods have their advantages and disadvantages and specific areas of application. Centralized control is the main method used in Industry 4.0, which results from the fact that production systems most often involve cyclical deterministic processes in which centralized control guarantees the achievement of the expected production effect within a specified time. It should be emphasized, however, that centralized control is effective in small production and transportation systems, especially in situations of lack of uncertainty. In the case of large industrial systems and those with stochastic elements, where centralized control is difficult, there is a need to look for other control methods. The literature contains many examples of research devoted to the decentralized control algorithms for large and distributed production and transport systems (Draganjac et al., 2016). The authors of these works emphasize the benefits of using the algorithms they have developed over the currently used centralized, hierarchical algorithms. Many researchers mention the need for decentralization (Fanti et al., 2018) considering the limitations of current centralized architectures as inappropriate to support particularly flexible production systems

FIGURE 3.3 Decentralized control.

within which nonstandard products are manufactured. The primary drawback of decentralized control lies in the heightened coordination required among system components, each pursuing individual objectives that may not result in an overall system optimum. The abovementioned feature is an argument in favor of searching for new centralized control algorithms.

The use of appropriate centralized control algorithms can lead not only to optimal results, but also to the scalability and flexibility of the control algorithm, which is a very desirable feature in the case of very large and complex manufacturing systems used in Industry 4.0. Centralized control is also more effective than decentralized control in large transport systems in which, during the implementation of transport tasks, stochastic situations occur that prevent the implementation of the transport task caused by, for example, congestion caused by other AGVs. The situation is similar in cases where we are dealing with hybrid systems in which there are AGVs and cooperating vehicles with human operators, for example, forklifts (Witczak et al., 2020a).

MODEL PREDICTIVE CONTROL ALGORITHM FOR AGV FLEET SCHEDULING

The production system in Industry 4.0 usually consists of machines, component and product storage stations, a transport system, and a manufacturing execution system (MES). The transport system might include AGVs and robotic loading stations. Assuming the production output delivers products sequentially, the MES characterizes them as: $\mathcal{P}(k)=\left[p(k),u(k),l(k),d(k)\right]$, where:

- $p(k) \in \{1,\dots,n_s\}$ denotes a transfer station identifier, with n_s indicating the number of such stations.
- $l(k)$ means the minimum time for loading the product from the output of the production system onto the AGV and its transport to the transfer station $p(k)$ from which the product is received to the warehouse.
- $u(k)$ represents the minimum time for unloading the product from the AGV to the transfer station $p(k)$ and returning the AGV to the production system.
- $d(k)$ denotes the minimum transport time for moving the $(k-1)$th product to transfer station $p(k-1)$ and the kth product to transfer station $p(k)$.

A succession of products progressing from production output to the transfer station $p(k)$ is expressed as

$$\mathcal{P}(0), \mathcal{P}(1), \dots, \mathcal{P}(n_{\mathcal{P}}-1), \tag{3.1}$$

where $n_{\mathcal{P}}$ represents the anticipated production timeline, and the kth loads are to be delivered to $p(k)$ at specified times:

$$x_r(0), x_r(1), \dots, x_r(n_{\mathcal{P}}-1). \tag{3.2}$$

In this scenario, the operation schedule for n_A AGVs must align with the product sequence at the system's output:

$$y(0), y(1), \dots, y(n_{\mathcal{P}}-1). \tag{3.3}$$

The efficiency of a transport system utilizing AGVs can be represented using the following cost function:

$$J(y) = -\sum_{k=0}^{n_{\mathcal{P}}-1} y(k). \tag{3.4}$$

Minimizing equation (3.4) is essential, subject to the scheduling constraint (3.2), while considering the performance of all n_A AGVs. A mathematical formulation of the AGV fleet is introduced to devise a real-time work plan using max-plus linear algebra (MPLA). Additionally, a MPC strategy is proposed to determine a sequence (3.3) that maximizes (3.4) under constraint (3.2). To represent the mathematical formulation of AGV fleet dynamics for prompt scheduling derivation, it is assumed that $x_i(k)$ indicates when the ith AGV is ready to transport the kth product, $x_{n_A+1}(k)$ signifies the arrival time of the kth product at $p(k)$, and $w_i(k) \in \{e, \varepsilon\}$ is a decision variable for the ith AGV regarding kth product transport, where $w_i(k) = e$ indicates task execution by the AGV and $w_i(k) = \varepsilon$ indicates an idle state. The state of $x_i(k)$ at given time intervals for an individual AGV can be expressed as:

$$x_i(k) = \max\left(x_i(k-1) + u_i(k-1) + l_i(k-1), y(k) + w_i(k)\right), \text{ for } i = 1, \dots, n_A, \tag{3.5}$$

where

$$u_i(k) = \max\left(e, u(k) + w_i(k)\right), \tag{3.6}$$

$$l_i(k) = \max\left(e, l(k) + w_i(k)\right), \tag{3.7}$$

$$w_i(k) = e \Leftrightarrow w_j(k) = \varepsilon, \quad \forall i \neq j. \tag{3.8}$$

The definition of the arrival time for the k th product to $p(k)$ transfer station is as follows:

$$x_{n_A+1}(k) = \max\left(x_1(k) + l(k) + w_1(k), \ldots, x_{n_A+1}(k-1) + d(k)\right). \tag{3.9}$$

The system characterized by (3.5)–(3.9) can be succinctly reformulated using discrete-event systems (DES) modeling and MPLA (Ribeiro et al., 2018). The classical MPLA restrictively defines a class of DES where only synchronization occurs. The proposed method bypasses these limitations using switching variables $\{(w_i(k), w_i(k-1))\}_{k=1}^{n_p}$.

Based on the $(max,+)$ structure of the algebra (R_{max}, ⊕, ⊗) introduced by (Butkovič, 2010), model (5)–(9) can be reformulated as:

$$x(k) = A\left(w(k-1), w(k), k\right) \otimes x(k-1) \oplus B\left(w(k), k\right) \otimes y(k) \tag{3.10}$$

while $x(k) = \left[x_1(k), x_2(k), \ldots, x_{n_A+1}(k)\right]^T$, $w(k) = \left[w_1(k), w_2(k), \ldots, w_{n_A}(k)\right]^T$, $A(.,.,.) \in \mathrm{R}_{max}^{n_{A+1} \times n_{A+1}}$, $B(.,.) \in \mathrm{R}_{max}^{n_{A+1}}$.

From (3.5) and (3.9), one can derive both $A\left(w(k-1), w(k), k\right)$:

$$\begin{bmatrix} u_1(k-1) + l_1(k-1) & \varepsilon & \cdots & \varepsilon \\ \varepsilon & u_2(k-1) + l_2(k-1) & \cdots & \varepsilon \\ \vdots & \vdots & \ddots & \vdots \\ u_1(k-1) + l_1(k-1) + l_1(k) + w_1(k) & u_2(k-1) + l_2(k-1) + l_2(k) + w_2(k) & \cdots & d(k) \end{bmatrix}$$

and $B\left(w(k), k\right)$:.

$$\left[w_1(k), w_2(k), \ldots, w_{n_A}(k), l(k)\right]^T.$$

For the reader's convenience, the matrices above can be denoted as $A_w(k)$ and $B_w(k)$.

The subsequent objective is to outline a method for estimating the product arrival time sequence (3.3) for a given n_p, ensuring minimization of (3.4) while considering (3.2) and the fleet efficiency of n_A AGVs. This method employs an MPC approach for MPLA systems, considering $w_i(k)$.

The main objective is to attain $y(k),\ldots,y(k+n_p-1)$ for $k,\ldots,k+n_p-1$, necessitating a refinement of (3.4) to the following form:

$$J(y) = -\sum_{j=0}^{n_p-1} y(k+j). \tag{3.12}$$

Building upon (Witczak et al., 2020a), it is possible to define a prediction vector $\tilde{w}(k) = \left[w(k)^T, w(k+1)^T, \ldots, w(k+n_p-1)^T\right]^T$ and corresponding vectors $\tilde{x}(k) = \left[x(k)^T, x(k+1)^T, \ldots, x(k+n_p-1)^T\right]^T$, $\tilde{y}(k) = \left[y(k)^T, y(k+1)^T, \ldots, y(k+n_p-1)^T\right]^T$. Applying (10) recursively results in:

$$\tilde{x}(k) = S(\tilde{w}(k)) \otimes \tilde{x}(k-1) \oplus T(\tilde{w}(k)) \otimes \tilde{y}(k). \tag{3.13}$$

where matrices $S(\tilde{w}(k))$ and $T(\tilde{w}(k))$ are derived via a series of recursive substitutions (refer to (Witczak et al., 2020a) for full details).

In addition to the scheduling (3.2), transportation (3.6)–(3.7), and concurrency (3.8) constraints, the MPC algorithm requires specification of the following production performance constraint:

$$y(k+1) - y(k) \geq y_z(k), \tag{3.14}$$

where $y_z(k) \geq 0$ denotes the maximum production efficiency.

Taking into account all constraints, the control approach for an AGV fleet is articulated through the following algorithm:

Algorithm: AGV fleet scheduling

***STEP* 0**:

Let $k = 1$, along with n_p and $w(0)$.

***STEP* 1**

Retrieve $\mathcal{P}(k),\ldots,\mathcal{P}(k+n_p-1)$, $y_z(k),\ldots,y_z(k+n_p-1)$, and $x_r(k), \ldots, x_r(k+n_p-1)$ from the MES.

***STEP* 2**

Considering (3.2), (3.6)–(3.8) as well as (3.14) compute $x(k-1)$:

$$\left(\tilde{y}(k)^*, \tilde{w}(k)^*\right) = \arg \min_{\tilde{y}(k), \tilde{w}(k)} J(y), \tag{3.15}$$

***STEP* 3**

Select the first components of vectors $\tilde{y}(k)^*$ and $\tilde{w}(k)^*$ for application in (3.5).

***STEP* 4**

Increment k by 1 and proceed to ***STEP* 1**.

CONCLUSIONS

This study examines fundamental challenges in scheduling and controlling AGVs within Industry 4.0 applications. In particular, the basic concepts related to the scheduling of tasks for AGVs operating in transport systems are presented. Moreover, the properties of centralized and decentralized control systems used in AGV fleets were analyzed. Moreover, the methods used in task scheduling for AGVs were reviewed. Furthermore, a new MPC algorithm for AGV fleet scheduling was also proposed. The algorithm is based on DES modeling and MPLA approaches. The advantage of the developed algorithm is its robustness to uncertainty that may potentially occur in the transport system. Due to its centralized nature, the proposed algorithm can be widely used in controlling fleets of AGVs in modern flexible manufacturing systems.

REFERENCES

Adams, J., Balas, E., & Zawack, D. (1988). The shifting bottleneck procedure for job shop scheduling. *Management Science*, *34*(3), 391–401.

Amirteimoori, A., Tirkolaee, E. B., Simic, V., & Weber, G. W. (2023). A parallel heuristic for hybrid job shop scheduling problem considering conflict-free AGV routing. *Swarm and Evolutionary Computation*, *79*, 101312.

Bocewicz, G., Bach, I., & Banaszak, Z. A. (2009). Logic-algebraic method based and constraints programming driven approach to AGVs scheduling. *International Journal of Intelligent Information and Database Systems*, *3*(1), 56–74.

Branisso, L. B., Kato, E. R. R., Pedrino, E. C., Morandin Jr, O., & Tsunaki, R. H. (2012). An intelligent autonomous vehicle management system. In *2012 Second Brazilian Conference on Critical Embedded Systems* (pp. 42–47). IEEE.

Butkovič, P. (2010). *Max-linear systems: theory and algorithms*. Springer Science & Business Media.

Chen, C., & Tiong, L. K. (2019). Using queuing theory and simulated annealing to design the facility layout in an AGV-based modular manufacturing system. *International Journal of Production Research*, *57*(17), 5538–5555.

Chen, J., Zhang, X., Peng, X., Xu, D., & Peng, J. (2022). Efficient routing for multi-AGV based on optimized Ant-agent. *Computers & Industrial Engineering*, *167*, 108042.

De Ryck, M., Versteyhe, M., & Shariatmadar, K. (2020). Methodology for a gradual migration from a centralized towards a decentralized control in AGV systems. In 2020 6th International Conference on Mechatronics and Robotics Engineering (ICMRE) (pp. 110–114). IEEE.

Draganjac, I., Miklić, D., Kovačić, Z., Vasiljević, G., & Bogdan, S. (2016). Decentralized control of multi-AGV systems in autonomous warehousing applications. *IEEE Transactions on Automation Science and Engineering*, *13*(4), 1433–1447.

Espinouse, M. L., Lacomme, P., Moukrim, A., & Tchernev, N. (2000). Branch and Bound Approach Based on Simulation for the Job Input Sequencing and Device Dispatching in FMS. *IFAC Proceedings* Volumes, *33*(17), 729–734.

Fanti, M. P., Mangini, A. M., Pedroncelli, G., & Ukovich, W. (2018). A decentralized control strategy for the coordination of AGV systems. *Control Engineering Practice*, *70*, 86–97.

Fatemi-Anaraki, S., Tavakkoli-Moghaddam, R., Foumani, M., & Vahedi-Nouri, B. (2023). Scheduling of multi-robot job shop systems in dynamic environments: mixed-integer linear programming and constraint programming approaches. *Omega*, *115*, 102770.

Fayazi, S. A., & Vahidi, A. (2018). Mixed-integer linear programming for optimal scheduling of autonomous vehicle intersection crossing. *IEEE Transactions on Intelligent Vehicles*, *3*(3), 287–299.

Gil-Gala, F. J., Mencía, C., Sierra, M. R., & Varela, R. (2020). Exhaustive search of priority rules for on-line scheduling. In *ECAI 2020* (pp. 2354–2361). IOS Press.

Ho, Y. C., & Liao, T. W. (2009). Zone design and control for vehicle collision prevention and load balancing in a zone control AGV system. *Computers & Industrial Engineering*, *56*(1), 417–432.

Hu, H., Jia, X., He, Q., Fu, S., & Liu, K. (2020). Deep reinforcement learning based AGVs real-time scheduling with mixed rule for flexible shop floor in industry 4.0. *Computers & Industrial Engineering*, *149*, 106749.

Jeong, I. J. (2024). A fully polynomial time approximation scheme for a stop schedule of a transportation line without branches. *Engineering Optimization*, 1–20.

Li, X., Wang, C., Hu, H., & Liang, Y. (2022). A polynomial time approximation scheme for the scheduling problem in the AGV system. In *2022 IEEE/RSJ International Conference on Intelligent Robots and Systems (IROS)* (pp. 11695–11702). IEEE.

Lipiec, B., Mrugalski, M., & Witczak, M. (2021). Health-aware fault-tolerant control of multiple cooperating autonomous vehicles. In 2021 IEEE International Conference on Fuzzy Systems (FUZZ-IEEE) (pp. 1–7). IEEE.

Luo, Y., Zhang, Q., & Lin, L. (2023). A Cooperative Hybrid Evolutionary Algorithm for Flexible Scheduling with AGVs. In 2023 International Conference on Sensing, Measurement & Data Analytics in the era of Artificial Intelligence (ICSMD) (pp. 1–7). IEEE.

Majdzik, P., Witczak, M., & Mrugalski, M. (2023). Scheduling the tasks of multiple AGVs in a fault-tolerant control way. *IFAC-PapersOnLine*, *56*(2), 150–155.

Ming, S. I., Bofan, W. U., Can, H. U., & Weiqiang, X.. (2024). Collaborative control method of intelligent warehouse traffic signal and multi-AGV path planning. *Journal of Computer Engineering & Applications*, *60*(11), 130–136.

Mrugalski, M. (2023). Identification and health-aware economic control of production systems: A fuzzy logic max plus algebraic approach. *Engineering Applications of Artificial Intelligence*, *120*, 105802.

Nageswara, R. M., Narayana, R. K., & Ranga, J. G. (2017). Integrated scheduling of machines and AGVs in FMS by using dispatching rules. *Journal of Production Eng.*, *20*(1), 75–84.

Novas, J. M., & Henning, G. P. (2014). Integrated scheduling of resource-constrained flexible manufacturing systems using constraint programming. *Expert Systems with Applications*, *41*(5), 2286–2299.

Riazi, S., Bengtsson, K., & Lennartson, B. (2020). Energy optimization of large-scale AGV systems. *IEEE Transactions on Automation Science and Engineering*, *18*(2), 638–649.

Ribeiro, G. R., Saldanha, R. R., & Maia, C. A. (2018). Analysis of decision stochastic discrete-event systems aggregating max-plus algebra and Markov chain. *Journal of Control, Automation and Electrical Systems*, *29*, 576–585.

Santos, J., Rebelo, P. M., Rocha, L. F., Costa, P., & Veiga, G. (2021). A* based routing and scheduling modules for multiple AGVs in an industrial scenario. *Robotics*, *10*(2), 72.

Seo, Y., Lee, C., & Moon, C. (2007). Tabu search algorithm for flexible flow path design of unidirectional automated-guided vehicle systems. *OR spectrum*, *29*, 471–487.

Shi, D., Mi, H., Collins, E. G., & Wu, J. (2020). An indoor low-cost and high-accuracy localization approach for AGVs. *IEEE Access*, *8*, 50085–50090.

Smutnicki, C., Pempera, J., Bocewicz, G., & Banaszak, Z. (2022). Cyclic flow-shop scheduling with no-wait constraints and missing operations. *European Journal of Operational Research*, *302*(1), 39–49.

Sun, P. Z., You, J., Qiu, S., Wu, E. Q., Xiong, P., Song, A., ... & Lu, T. (2022). AGV-based vehicle transportation in automated container terminals: A survey. *IEEE Transactions on Intelligent Transportation Systems*, *24*(1), 341–356.

Sze, J. F., Salhi, S., & Wassan, N. (2016). A hybridisation of adaptive variable neighbourhood search and large neighbourhood search: Application to the vehicle routing problem. *Expert Systems with Applications*, *65*, 383–397.

Vivaldini, K. C., Rocha, L. F., Becker, M., & Moreira, A. P. (2015). Comprehensive review of the dispatching, scheduling and routing of AGVs. In CONTROLO'2014–proceedings of the 11th Portuguese conference on automatic control (pp. 505–514). Springer International Publishing.

Vlachos, I. P., Pascazzi, R. M., Zobolas, G., Repoussis, P., & Giannakis, M. (2023). Lean manufacturing systems in the area of Industry 4.0: A lean automation plan of AGVs/IoT integration. *Production Planning & Control*, *34*(4), 345–358.

Wang, Z., & Zeng, Q. (2022). A branch-and-bound approach for AGV dispatching and routing problems in automated container terminals. *Computers & Industrial Engineering*, *166*, 107968.

Wang, X., Wu, W., Xing, Z., Chen, X., Zhang, T., & Niu, H. (2022b). A neural network based multi-state scheduling algorithm for multi-AGV system in FMS. *Journal of Manufacturing Systems*, *64*, 344–355.

Witczak, M., Lipiec, B., Mrugalski, M., Seybold, L., & Banaszak, Z. (2020a). Fuzzy modelling and robust fault-tolerant scheduling of cooperating forklifts. In 2020 IEEE International Conference on Fuzzy Systems (FUZZ-IEEE) (pp. 1–10). IEEE.

Witczak, M., Mrugalski, M., Pazera, M., & Kukurowski, N. (2020b). Fault diagnosis of an automated guided vehicle with torque and motion forces estimation: A case study. *ISA Transactions*, *104*, 370–381.

Witczak, M., Seybold, L., Lipiec, B., Aubrun, C., & Mrugalski, M. (2022). An economic fault-tolerant transportation strategy for multiple assembly station systems. *IFAC-PapersOnLine*, *55*(10), 1716–1721.

Xiang, J., Zhang, Y., Cao, X., & Zhou, Z. (2023). An improved multi-objective hybrid genetic-simulated annealing algorithm for AGV scheduling under composite operation mode. *CMC-Computers, Materials & Continua*, *77*(3), 3443–3466.

Xu, B., Fan, H., & Li, J. (2024). Energy-efficiency and safety-driven multidimensional evolutionary game for AGVs transportation at automated container terminals. *Computers & Industrial Engineering*, *192*, 110192.

Yuan, M., Zheng, L., Huang, H., Zhou, K., Pei, F., & Gu, W. (2023). Research on flexible job shop scheduling problem with AGV using double DQN. *Journal of Intelligent Manufacturing*, *36*(1), 509–535.

Zacharia, P. T., & Xidias, E. K. (2020). AGV routing and motion planning in a flexible manufacturing system using a fuzzy-based genetic algorithm. *International Journal of Advanced Manufacturing Technology*, *109*(7), 1801–1813.

Zhang, F., Li, R., & Gong, W. (2024). Deep reinforcement learning-based memetic algorithm for energy-aware flexible job shop scheduling with multi-AGV. *Computers & Industrial Engineering*, *189*, 109917.

Zhang, L., Hu, X., & Song, T. (2023a). Design and optimization of AGV scheduling algorithm in logistics system based on convolutional neural network. In *2023 2nd International Conference on Artificial Intelligence and Autonomous Robot Systems (AIARS)* (pp. 167–171). IEEE.

Zhang, Z., Chen, J., & Guo, Q. (2023b). Application of automated guided vehicles in smart automated warehouse systems: A survey. *CMES-Computer Modeling in Engineering & Sciences*, *134*(3), 15–29.

Zheng, Y., Xiao, Y., & Seo, Y. (2014). A tabu search algorithm for simultaneous machine/AGV scheduling problem. *International Journal of Production Research*, *52*(19), 5748–5763.

Zou, W. Q., Pan, Q. K., Wang, L., Miao, Z. H., & Peng, C. (2022). Efficient multiobjective optimization for an AGV energy-efficient scheduling problem with release time. *Knowledge-Based Systems*, *242*, 108334.

4 Extended Interdependency Matrix to Evaluate Influence Factors in Circular Value Creation Systems

Application in Industry

Nada Ruzicic, Anja Braun, and Peter Ohlhausen

INTRODUCTION

Industry 5.0 builds upon the technological advancements of its predecessor, incorporates principles of the circular economy (CE), adopts a human-centric approach to industrial development, and prioritises sustainability as the cornerstone of a resilient economy (European Commission, 2023). The state of the Earth's system is evaluated through nine critical planetary boundaries, which define the safe operating limits for human activities. Six of these boundaries have already been breached as a result of human activities (Richardson et al., 2023). Transitioning towards greater sustainability requires an increased focus on optimising production systems, fostering collaborative relationships, and developing environmentally sustainable logistics (Bag et al., 2021). The CE is increasingly regarded as a key solution to the challenges of climate change and resource scarcity.

Integrating circular processes into existing value creation systems is a multifaceted challenge due to the interconnections inherent in the CE (Saidani et al., 2017; Kirchherr et al., 2023; Suzanne et al., 2020).

Transitioning to circular value creation systems is particularly demanding because many products are not designed for circularity, and numerous interrelated influence factors must be considered. These factors are often interconnected, further complicating the shift towards a CE. This paper aims to address the research gap regarding influence factors in the CE by proposing a solution-oriented approach using an interdependency matrix. In continuation of the study by Ruzicic et al. (2024), this paper provides extended examples and details of the expansion and optimisation of the developed interdependency matrix. Compared to the work of Lee et al. (2024), which

DOI: 10.1201/9781003625360-4

focuses extensively on CE indicators, this study emphasises the production perspective and adopts a correspondingly higher level of abstraction to reduce complexity.

While the results are not comprehensive, the interdependency matrix is designed as a dynamic tool that undergoes continuous improvement and adaptation. It provides an initial framework for researchers and practitioners seeking to explore and apply CE principles.

RESEARCH DESIGN

The research design employed in this study adopts a mixed-methods approach, as outlined by Schoonenboom and Johnson (2017). To establish the state of the art, a literature review was conducted, primarily utilising ScienceDirect, supplemented by searches on Google Scholar. The initial focus was on identifying existing metrics and evaluation tools for assessing the circularity of companies, which was later broadened to include additional circularity indicators and requirements. This research was further expanded to structure key influence factors relevant to designing circular value creation systems as described in Ruzicic et al. (2024).

The insights gathered from the literature review are enriched with expertise from professionals in industry and academia, many of whom are part of the Cyclometric project network. The expert interviews were guided by an initial draft of the interdependency matrix and an accompanying overview connecting factors and indicators for the analysis of interdependencies. These outcomes were then integrated into the iterative development of the interdependency matrix.

As displayed by Ruzicic et al. (2024), relationships between individual influence factors were assessed using distinctive morphological analyses. A case study was employed to provide a reference point and framework for applying the developed interdependency matrix.

To refine and validate the matrix, a verification process was undertaken, generating application examples based on the case study. These examples were designed as practical tests to evaluate the matrix's usability and robustness in real-world scenarios.

STATE OF THE ART AND SCIENCE

In the context of circular value creation, this paper draws upon the Cradle-to-Cradle framework introduced by McDonough and Braungart (2002). This study adopts the definition of the CE provided by Kirchherr et al. (2023).

> The circular economy is a regenerative economic system [...] with the aim to promote value maintenance and sustainable development, creating environmental quality, economic development, and social equity, to the benefit of current and future generations. [...].
>
> *Kirchherr et al. (2023)*

This study also builds on the nine strategies for advancing a CE (R-strategies) as outlined by Potting et al. (2017). These strategies distinguish and evaluate the effects of different approaches to product handling, such as remanufacturing, repair, and reuse, and are collectively referred to as "R-strategies". Consistent with Potting et al. (2017), the term "R-" is used in this paper to outline these nine strategies for the CE.

The CE offers a framework for fundamentally transforming current business models towards preventive and regenerative eco-industrial development (Ghisellini et al., 2016). According to Ghisellini et al. (2016), greater commitment and additional measures towards the CE are required. Circular production and logistics necessitate a revised structure that emphasises increased and more frequent cooperation with both suppliers and customers (Schmitt et al., 2021; Zucchella and Previtali, 2019; Baldassarre et al., 2020). Production planning in a CE should integrate considerations for additional process steps, supplementary space, and extra time required to manage the diagnosis of those returned items and their reintegration into the value creation system (Schmitt et al., 2021).

Regulated markets and the required infrastructure for the widespread implementation of remanufacturing and reuse are currently insufficient (Kara et al., 2022). In both reducing pollution and waste, as well as enhancing productivity, circular manufacturing plays a key role (Nascimento et al., 2019).

Companies need concepts for establishing circular value creation structures (Rossi et al., 2020). Information and material flows aligned with the CE disrupt the traditional structure of well-established decision-making rooted in a linear production model. To address this shift, reverse logistics must be integrated into and accounted for within production environments (Suzanne et al., 2020).

Although existing tools and methods offer initial insights into the circularity of products, the CE is still far from being fully integrated in its complexity (Saidani et al., 2017). By combining various methods, it becomes possible to better understand the interrelationships and address potential conflicts between sustainability goals, including CE and waste management principles. Additionally, diverse tools can be applied to clarify conflicting objectives and identify risks in order to prevent unintended negative outcomes (Meidl, 2021). According to Lee et al. (2024) and Elia et al. (2017), there is a need to simplify and improve the evaluation methods within the context of CE.

To address the complexity of interrelationships, this research paper explores how the interdependencies of influence factors for circular value creation systems can be evaluated.

DEVELOPMENT OF THE CIRCULAR VALUE CREATION INFLUENCE FACTORS INTERDEPENDENCY MATRIX

Evaluating CE using various existing metrics involves a high level of complexity (Saidani et al., 2017). It is important to note that many existing indicators and metrics frequently address impacts not only at the process level but also at broader scales, such as entire companies (including production and office activities), countries or regions, and even the global level (OECD 2020). As highlighted by Ruzicic et al. (2024), certain metrics and indicators tend to focus on specific materials and scenarios, rather than offering a comprehensive and holistic view of production.

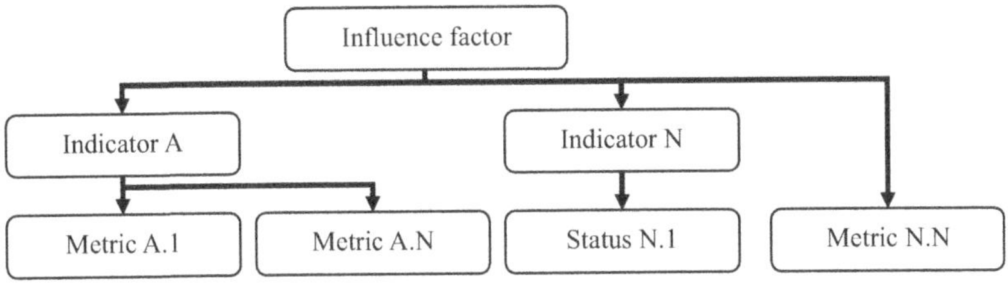

FIGURE 4.1 Visualisation of the relationship between factors, indicators, and metrics. (Source: Ruzicic et al., 2024.)

To evaluate production metrics effectively, it is vital to have a thorough understanding of material compositions in products, process structures, and transportation methods. Such information and real-world data on a company's value creation processes are often not entirely accessible, particularly during the production planning phase. As a result, it takes a production-centric and value creation approach, setting it apart from existing research on CE assessments as described by Ruzicic et al. (2024).

Users can also customise and extend the interdependency matrix as an evaluation framework, adapting it to their specific needs, company data, and real-world scenarios (Ruzicic et al., 2024).

Figure 4.1 is extracted from Ruzicic et al. (2024). This diagram shows the relationship between indicators and influence factors. These factors may include relevant aspects, requirements, and other essential elements for a circular value creation system. An influence factor may encompass one or more indicators and metrics, as explained by Ruzicic et al. (2024). According to the understanding of the present study, an indicator provides either a quantitative or qualitative value that reflects the status of a defined aspect. This can be represented through a status (whether something is the case or not) or metrics (numerical values). In some cases, a metric is not included in an indicator but is instead directly associated with an influence factor, where it is used for its evaluation (Ruzicic et al., 2024).

Ruzicic et al. (2024) stated that including metrics in the interdependency matrix would introduce too much complexity at the current stage of the work. Therefore, metrics were excluded from the matrix. Metrics are presented in a second table, linked to the interdependency matrix, within the indicators section. A single influence factor can have multiple associated indicators as outlined by Ruzicic et al. (2024).

A single influence factor can have multiple associated indicators. For instance, various indicators might measure energy consumption at different stages of a production system. Qualitative indicators or metrics can also be added to influence factors by institutions or through further research (Ruzicic et al., 2024). The microlevel aspects are illustrated with a few examples and can be further developed through future research efforts (Ruzicic et al., 2024).

Table 4.1 has been adapted with improved wording from Ruzicic et al. (2024). It illustrates how the influence factors are categorised and subcategorised in relation to the design of circular value creation systems. As outlined by Ruzicic et al. (2024),

TABLE 4.1
Influence factors for circular value creation systems ordered by categories and subcategories

Category	Subcategory	Influence factors
Product design	Material-specific properties	Material-related emissions, toxicity, renewable raw material, wear properties of the material, R-promoting reprocessing properties of the material
	Design-specific processing and repairability	Modularity of the product, wear properties of the product (number of wear parts, design-related abrasion, etc.), R-strategy friendly joining techniques, reusability of components, repairability
Business model and corporate strategy	Utilisation	Product lifespan
	Take-back system	Accessibility to take-back system, reward system for take-back system
	Costs	Margin per R-step
Material flow	Material	Circularity of the material flow
Production	Energy in processes	Process energy, energy in support processes
	Energy in transports	Upstream transports, downstream transports, energy consumption of reverse logistics transports
	Heat energy	Utilised waste heat (process input), reutilised waste heat (process output)
	Production waste	Material waste (packaging, drill cuttings), faulty materials and goods
	Production emissions	Air pollution, water pollution, soil pollution, process emissions
	Possibility of recirculation (reprocessing)	Recirculation of water, material, and energy (heat)
	Costs	Work in process (WIP) costs, material costs, and cost savings through the use of secondary material
Reprocessing	R-strategy reprocessing	Percentage of reprocessed material

Source: Adapted from Ruzicic et al. (2024).

the categories are based on adaptations from Aranda-Usón et al. (2020), Saidani et al. (2017), and the OECD report on CE in cities and regions (OECD, 2020). The categories have been tailored to align with circular value creation systems.

Ruzicic et al. (2024) explained that influence factors can inherently be positive, negative, or neutral in nature. For instance, emission-related influence factors are regarded as detrimental to the development towards a CE. "Renewable raw material" and "Modularity" are deemed positive influence factors. Whether the influence factor

"Margin for R-processed" can be seen as positive or negative relies on the value of the influence factor. To enhance comparability and improve the clarity of the interdependency matrix, Ruzicic et al. (2024) expressed the influence factors within it in terms of their supportive impact on the CE. Therefore, terms such as "reduction", "minimisation", and "R-promoting" have been added to the names of the influence factors (Ruzicic et al., 2024). As defined by Ruzicic et al. (2024), if the influence factors are beneficial for the CE, the background of the label in the interdependency matrix is green. It turns red if the term suggests a reduction. Neutral impact factors are represented by the colour blue. These neutral influence factors derive their negative or positive tendency from the structure and situation of the individual value creation system under consideration (see Figure 4.2).

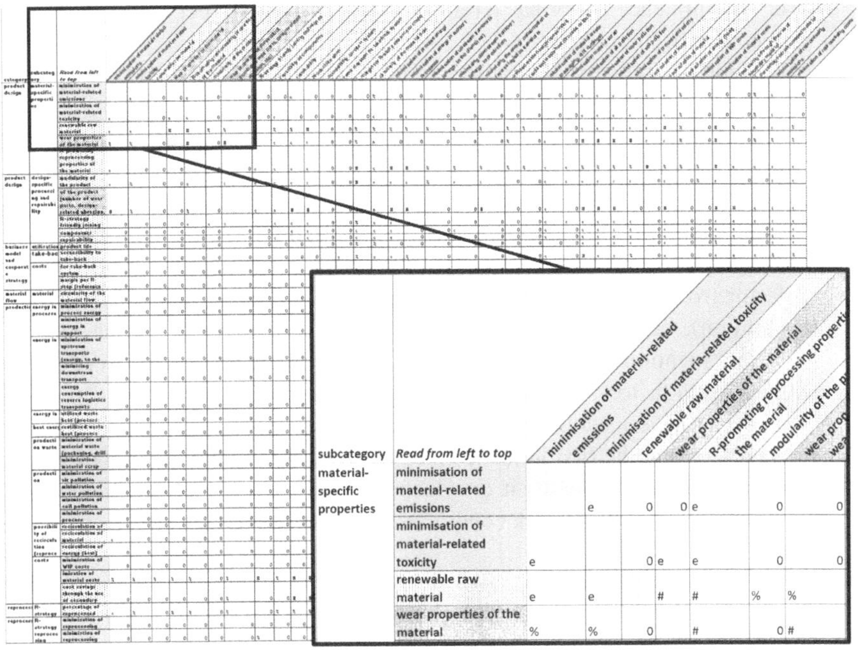

Influence factors that are:

positive by their nature (originally highlighted in *green*)

neutral by their nature (originally highlighted in *blue*)

negative by their nature and received a positive orientation through wording (originally highlighted in *red*)

Cells where the same influence factor is found on both the vertical and horizontal axes and therefore contain no entries for defined interdependencies

Zoom into the Circular Value Creation Influence Factors Interdependency Matrix

FIGURE 4.2 Structure of the circular value creation influence factors interdependency matrix with a zoom-in section. (Adapted from Ruzicic et al., 2024.)

The influence factors are set horizontally and vertically in the developed two-dimensional and multilayer interdependency matrix as shown in Figure 4.2. At this time, the interdependency matrix is presented in an Excel spreadsheet. The reading direction of the matrix is from left to right. Another spreadsheet displays a comparison of associated indicators and metrics with the influence factors. In addition, explanations for terms, symbols, and colours used are documented in the Excel sheets to prevent misinterpretation. On one of the worksheets in this Excel document, the parallels and interfaces with other research institutions and companies are noted.

As part of the Cyclometric project, a transfer of the results into a software tool is planned.

The interdependencies are represented by symbols as defined in the previous work of Ruzicic et al. (2024) and quoted verbatim in the following:

"[...]

- e: The influence factor has a reinforcing, increasing effect on the other influence factor.
- 0: Neutral – influence factor has no influence, no connection to the influence factor.
- n: The influence factor has a reducing, softening effect on the other influence factor.
- %: It may or may not have an influence (case-specific); whether it has an influence depends on the processes, the product, and the framework conditions.
- &: It has an indirect effect (via other factors or aspects); one influence factor (row) does not have a direct influence on the second influence factor (column), but by certain constraints (it does not have to be the influence factors mentioned), as these constraints, in turn, create an indirect relation to the second factor.
- #: It can have a reinforcing or reducing effect (system/process-dependent); it is clear that the first influence factor affects the second factor, but whether it increases or decreases it or whether the influence is positive or negative depends on the framework conditions/processes/materials, etc." Ruzicic et al. (2024).

FINDINGS

Ruzicic et al. (2024) stated that the interdependencies between two factors can vary. In case of toxicity and material, toxicity has no impact on the material itself, but the use of renewable materials can influence toxicity. The findings of Ruzicic et al. (2024) illustrate how the individual influence factors affect each other.

Influence factors can have a positive (increasing), negative (decreasing), or situation-dependent impact on one another. In some cases, the processes, product designs, and company structures will determine whether certain influence factors have interdependencies (Ruzicic et al., 2024). Some influence factors can have an indirect influence on each other through other influence factors or aspects. It is difficult to determine whether the influence of one factor on another is strong or not (Ruzicic et al., 2024).

The research findings of Ruzicic et al. (2024) indicated that while some influence factors are important to consider, they could distort the evaluation of circular production and business. As a result, these factors were excluded from the interdependency matrix. One of the excluded influence factors is the "Circular economy culture" (Ruzicic et al., 2024). Investment costs are also not initially suitable to be considered in the interdependency matrix, as further explained by Ruzicic et al. (2024).

Some categories are more interdependent than others. The findings of Ruzicic et al. (2024) revealed that influence factors associated with "Reprocessing" are closely linked to those in the "Business Model and Corporate Strategy" category. The category "Design" has a significant impact on the factors related to "Reprocessing", "Production", and "Material Flow". The research behind the interdependency matrix indicates that the factors often have a mutually positive influence.

It should be noted that the matrix shows that reducing costs and promoting a CE are not contradictory objectives, but can belong together. An example of a negative influence of one factor on another is the interdependence of the circularity of material flows on the minimisation of energy demand for transportation in reverse logistics. The explanation behind this interdependency is that increasing the circularity of material flows requires more transportation in reverse logistics, which in turn demands more energy. One measure to counteract this negative effect would be the use of alternative transportation modes and vehicles that require less energy.

The interdependency matrix indicates that there are few trade-offs between the influence factors of a circular value creation system. Although the interdependency matrix does not claim to be exhaustive, only nine of the analysed interdependencies negatively correlate with each other. This corresponds to under one per cent of the 1332 analysed interdependencies between the influence factors.

The first version of the interdependency matrix, presented by Ruzicic et al. (2024), revealed the complexity of the factors to be considered when transitioning to circular value creation systems. One reason for this increased complexity is that it is not always possible to predict whether a change in the value of one influence will positively or negatively affect another or the system as a whole (Ruzicic et al., 2024).

VALIDATION IN THE LEARNING FACTORY WERK150 BASED ON THE USE CASE

As outlined in Ruzicic et al. (2024), creating application examples for a case study is a key part of the matrix validation.

A learning factory is an educational environment that authentically simulates processes, includes multiple stations, and integrates both organisational and technical elements (Abele et al., 2017). The structure of this environment is flexible and models a real value chain (Abele et al., 2017). Institutions such as learning factories can provide a realistic testing environment for the testing and development of new methods and tools in the production environment, as well as represent practical examples for industry (Koller et al., 2023; Abele et al., 2017). Knowledge transfer in learning factories is a key enabler of the transition towards a CE in industry (Kadner et al., 2021).

Based on the infrastructure and production lines of the "Werk150" learning factory at the Reutlingen University campus, the use case is developed with the help of

Category	Subcategory	*Read from left to top*	Minimisation of material-related emissions	Explanatory examples	Minimisation of materia-related toxicity	Explanatory examples	Renewable raw material	Explanato
Product design	Material-specific properties	Minimisation of material-related emissions			e		0	
		Minimisation of material-related toxicity	e				0	
		Renewable raw material	e		e			
		Wear properties of the material	%		%		0	
		R-promoting reprocessing properties of the material	e		0		0	

Influence factors that are:

- Positive by their nature (originally highlighted in *green*)
- Neutral by their nature (originally highlighted in *blue*)
- Negative by their nature and received a positive orientation through wording (originally highlighted in *red*)

- Cells without where the same influence factor is found on both the vertical and horizontal are axes and therefore contain no entries for defines interdependencies
- Schematically highlighted cells where application examples are noted (originally highlighted in *purple*)

FIGURE 4.3 Zoom into the circular value creation influence factors interdependency matrix showing schematically the space for the examples of validation. (Adapted from Ruzicic et al., 2024.)

morphological analysis (Ruzicic et al., 2024). This use case is employed to create application examples, which are used to test the robustness and applicability of the developed independency matrix (Ruzicic et al., 2024).

In Figure 4.3, the application examples are schematically highlighted in purple and placed next to the estimation symbols on the right side of the matrix.

Building on the previous work of Ruzicic et al. (2024), the validation was expanded and further refined through examples and explanations of the assessed interdependencies.

In the following, examples are given for each type of defined interdependency.

Examples for [e] – "The Influence Factor Has a Reinforcing, Increasing Effect on the Other Influence Factor"

Influence factor "Modularity of the product" on the influence factor "Repairability".

Thanks to the modular design, defective parts can be easily identified and replaced without having to dismantle the entire product. This speeds up the repair process considerably and reduces product downtime.

> Influence factor "Wear properties of the product (number of wear parts, design-related abrasion, etc.)" on the influence factor "R-strategy friendly joining techniques".

If the product is subject to heavy wear, the materials used for the joining technology should be wear-resistant and durable. High-quality materials such as stainless steel or special alloys can increase the durability of the fasteners and extend their service life.

> Influence factor "Minimisation of toxicity" on the influence factor "Wear properties of the material".

Abrasion produces particles that can have an environmentally harmful effect on metals and plastics. Internalisation of toxicity therefore means a reduction in wear properties. Depending on whether it is a protective wear component made of sustainable material or whether the abrasion is undesirable. In this case, the effect on the reduction in toxicity varies. In general, however, the effect on the wear properties is positive. Because if the design is changed to a renewable raw material, more abrasion is also environmentally friendly.

Examples for [0] – "Neutral – Influence Factor Has No Influence, No Connection to the Influence Factor"

> Influence factor "Recirculation of energy (heat)" on the influence factor "Minimisation material waste".

Waste heat reutilised from the production processes has no influence scrap rate of the production.

> Influence factor "Minimisation of air pollution" on the influence factor "Minimisation of material-related emissions".

This interdependence shows that if there is an effect of influence factor A on influence factor B, it does not imply the reverse effect from factor B back onto A. While the reduction of material-related emissions has an impact on minimising air pollution, air pollution does not, in turn, affect the reduction of material-related emissions.

Example for [N] – "The Influence Factor Has a Reducing, Mitigating Effect to the Other Influence Factor"

> Influence factor "Wear properties of the material" on the influence factor "Minimisation of material-related emissions.

When material (characteristic of the component) wears out, emissions are usually produced in some form, e.g. through abrasion. Even if it is a natural material, it may have to be replaced/re-produced and production then causes emissions.

The formulation for the factors of the interdependency matrix was chosen so that they all promote the CE. This means that the cost-related factors are formulated in such a way that they result in savings. For example, minimising the cost of work in progress and cost savings through the use of secondary materials.

The formulation emphasises cost reduction. With other wording, e.g. a single word "costs", the user might think of expensive or high costs, as costs generally have a negative connotation. The different emphasis would be confusing. This would lead to the interdependency between cost-related influence factors and other influence factors being almost always negative, while not comparable.

Examples for [%] – "It May or May Not Have an Influence"

Influence factor "Wear properties of the material" on the influence factor "R-promoting reprocessing properties of the material".

If a product is subjected to heavy use, its particle size and shape can change. Wear and tear caused by use can change the surface properties of products, which in turn affects the efficiency of reprocessing.

Influence factor "Accessibility to take-back system" on the influence factor "Margin per R-step".

An easily accessible take-back system that offers consumers an easy way to return products can increase the return rate. This means that more material is available for recycling, which increases the efficiency of the recycling process and potentially reduces costs.

Influence factor "Percentage of reprocessed material" on the influence factor "Wear properties of the material".

Reprocessing processes can alter the material properties. Treatment with heat can, for example, change the material structure of plastics. Recycled material may contain impurities that affect the homogeneity of the material. Impurities and altered material structures can serve as points of failure, increase wear, and decrease the lifespan of a component or product.

Influence factor "Minimisation of reprocessing emissions" on the influence factor "Recirculation of water".

These two influence factors provide an example that the interdependencies are reliant on the company structure and internal processes. The influence factor *Minimisation of reprocessing emissions* may have an impact on the water used in the production process or on its return in production. The influence depends, on the one hand, on whether water is used in production and processing. On the other hand, the effect depends on whether production and processing are linked (e.g. bidirectional vs. completely separated processes). If both conditions are met in a particular case, the defined interdependency "it may or may not have an influence" with the symbol of [%] can be transferred to "the influence factor has a reinforcing effect" with an [e]. Companies must examine this on a case-by-case basis for themselves.

Examples for [&] – "It Has an Indirect Effect"

> Influence factor "Reutilised waste heat (process output)" on the influence factor "Margin per R-step".

The utilisation of reused waste heat can help to improve the overall efficiency of operations. If waste heat is utilised effectively, it can help production processes run more smoothly, reduce bottlenecks, and increase operational efficiency, which in turn reduces costs.

> Influence factor "Margin per R-step" on the influence factor "Circularity of the material flow"

A higher margin for recycled materials compared to virgin raw materials can motivate companies to collect, sort, and reuse or recycle recyclable materials. This leads to a more efficient material flow, as fewer new resources are required.

> Influence factor "Accessibility to take-back systems" on the influence factor "Minimisation of process energy".

Take-back systems return products and materials to the production cycle instead of disposing of them. This promotes a CE in which resources are utilised more efficiently. By reusing or recycling products, the need for new raw materials can be reduced, which in turn minimises energy consumption in the processing of raw materials. In this way, the accessibility of the take-back system can have an indirect influence on the minimising process energy.

Examples for [#] – "It Can Have a Reinforcing or Reducing Effect (System/Process-dependent)"

> Influence factor "Minimising material costs" on the influence factor "Repairability".

As secondary materials can come from recycled sources, they are often more readily available than new materials. This can increase the availability of spare parts and speed up repairs. Secondary materials can be of lower quality than new materials. If inferior materials are used, this can reduce the life of the product and affect repairability.

> Influence factor "Reward system for take-back system" on the influence factor "Minimisation of water pollution".

Incentives such as deposit or take-back systems motivate consumers to return their used products instead of disposing of them through household waste or landfills. This increases the recycling rate and reduces the amount of waste that could enter waterways. Repairable components can be processed and reused. As a result, there is less need to produce new components, and less water is polluted through the production of components. Depending on the product (e.g. tyres or engines) and the reprocessing, the associated cleaning process can be polluting. Here, the quantity, type of product, processes, and structures determine whether the effect is positive or negative and which measures a company should take to foster a circular value creation system.

The influence factors "Wear properties of the material" and "Wear properties of the product (number of wear parts, design-related abrasion, etc.)" show many elements with the symbol [#]. This is because, depending on the product and component, either minimal wear is sustainable, or wear can be intentional and sustainable. The latter is the case, for example, when a component serves to protect the overall product and is easily replaceable. Alternatively, it may be a sustainable material that experiences greater wear but is still harmless to the environment.

The validation showed that the interdependency matrix can be used in the application context. The interdependency matrix was effectively applied to the case study, providing a strong basis for discussion. Various possibilities were highlighted in the example column, making it clear why a particular estimation was made. The interdependency matrix does not claim to be comprehensive. The matrix must be critically examined in a case-specific manner. This very questioning leads to conclusions that can be used to further develop and customise the matrix. The interdependency matrix provides a foundation for examining the interdependencies of the respective feature characteristics for each application case. From this, decisions and actions can be derived to further develop a company towards a CE.

EXPANSION AND OPTIMISATION OF THE CIRCULAR VALUE CREATION INFLUENCE FACTORS INTERDEPENDENCY MATRIX FOR APPLICATION IN THE INDUSTRY

To better visualise where there are lots of correlations between factors and which factors are particularly relevant, the correlations were summed up. For the total values, reducing and softening effects [n] were deducted from reinforcing, increasing effects [e].

The defined interdependency "it may or may not have an influence" with the symbol [%] is calculated as zero, just like "no influence". The reason for this is that the existence of an influence is case-specific, and it is not possible to predict how likely an impact will occur. When applying the interdependency matrix to a specific case, users of the matrix should adjust the calculated value for the defined interdependency "it may or may not have an influence". They can then convert it to "the influence factor has a reinforcing, increasing effect on the other influence factor", "influence factor has no influence, no connection to the influence factor", or "the influence factor has a reducing, softening effect to the other influence factor".

After validation, it was found that the defined interdependency "it has an indirect influence" with the symbol [&] is difficult to apply when using the interdependency matrix. The reason for the difficulty is that the indirect influence is very unspecific, depends on unavailable background information, and has an unclear probability. Taking into account not only structural and design aspects of a company's value creation system (reverse logistics, distances to suppliers, possibility to handle r-processes internally), but also possible indirect influence chains complicates the interdependency matrix, which can lead to less qualitative results. The uncertainties of the defined interdependency "it has an indirect influence" were difficult to capture when determining the influence factors with the highest impact. The defined interdependency "it has an indirect influence" was therefore given a lower weighting when analysing the influence factors. The validation process showed that some of the defined interdependencies "it has an indirect influence" [&] can also be mapped by the other influence factors if necessary.

An example of an analysed interdependency indicated by the designated symbol of [&], mapped by another influence factor, is the influence factor "Reutilised waste heat (process output)" on the influence factor "Margin per R-step". The examined interdependency is associated with energy used in processes. This interdependency is also covered by "Minimisation of process energy" on "Margin per R-step" and therefore set on "no influence" [0].

The defined interdependency of "Margin per R-step" on "Circularity of the material flow" has been transformed into "the influence factor has a reinforcing, increasing effect" [e], as the interdependency is dominant and should not be neglected. If the margin per R-step is too low or negative due to high reprocessing costs, this can compromise the competitiveness of companies. Companies would then decide to reprocess fewer products, components, and materials, which would decline the circularity of the material flow.

The third example of the defined interdependence of "Accessibility to take-back systems" on "Minimisation of process energy" has also been transferred to "the influence factor has a reinforcing, increasing effect" [e]. The justification for this change is based on the hypothesis that through the return of products and components, fewer new components need to be produced, as old components can be processed and then reused. By avoiding new productions, process energy in manufacturing can be reduced.

Therefore, the interdependency "it has indirect influence" [&] was transferred to "has no influence" [0], "the influence factor has a reinforcing" [e] or to "it may or may not have an influence" [%]. This has increased the total number of factors that do not influence each other or may have an influence depending on circumstances. The decision as to which of the options was chosen depended on which of the possibilities was more obvious and seemed realistic based on the examples.

The interdependency defined as "it can have a reinforcing or reducing effect" with the symbol [#] indicates that there is a definite influence between two factors in individual cases. This influence can be either reinforcing or reducing, depending on the product and corporate structure. Consequently, this defined interdependency creates a variance.

Whether the effect is reinforcing or reducing cannot be defined in general terms and must be individually examined by the user on a case-by-case basis. The examples and explanations listed in the interdependency matrix describe in which cases there can be a reinforcing effect and when there can be a reducing effect.

When considering a row in the interdependency matrix with multiple symbols [#], there are three options: The first option is that all elements with the symbol [#] represent a reinforcing effect; the second option is that all symbols [#] represent a reducing effect; and the third option is that some elements indicate a reinforcing effect while others indicate a reducing effect. This variance leads to the totals not

Category	Subcategory	*Read from left to top*	Minimistion of reprocessing costs	Explanatory examples	Positive influences [e]	Negative influences [n]	Positive-or-negative influences [#]	Maximum: sum with all positive #	Minimum: sum with all negative #	Average main value
Product design	Design-specific processing and repairability	Modularity of the product	e	By replaci	22	0	1	23	21	22
		Wear properties of the product (number of wear parts, design-related abrasion, etc.)	e	Products v	5	− 2	− 12	15	= -9	3
		R-strategy friendly joining techniques	e	Screw con	21	0	0	21	21	21
		Reusability of components	e	Instead of	16	0	0	16	16	16
		Repairability	e	By extendi	15	0	0	15	15	15

Meaning of the symbols used for visualization of the calculation steps:

- ✚ Plus symbolising the next number/column is added
- ▬ Minuses symbolising the next number/column is subtracted
- = The equals sign symbolises that the result of the outlined calculation steps is in the column
- ⌣ Arrows for the calculation of the maximum
- ⌢ Arrows for the calculation of the minimum
- □ Frames of the minimum and maximum value from which the average is calculated (calculation: sum of the maximum and minimum divided by two)
- ⇨ Arrow pointing from minimum and maximum to the average main value

Influence factors that are:

- Positive by their nature (originally highlighted in *green*)
- Neutral by their nature (originally highlighted in *blue*)
- Negative by their nature and received a positive orientation through wording (originally highlighted in *red*)

- Dividing lines between the areas of the matrix (overview of interdependencies and examples, and calculative values)

- Highlighting in various shades of green (here in light grey and white) for high values of the numerical range calculated
- Highlighting in yellow (here in medium dark grey) for medium values of the numerical range calculated

FIGURE 4.4 Segment of the circular value creation influence factors interdependency matrix with horizontal sums and average main values.

having a fixed value, but rather a range between a minimum and maximum value (see Figure 4.4 and Figure 4.5).

For the calculation, the sum of the reinforcing effects [e] was added, while the sum of the reducing effects [n] was subtracted. For the minimum value, it was assumed that elements in the interdependency matrix with the symbol [#] have a reducing effect, and this sum was subtracted from the value of all interdependencies with a reinforcing effect [e]. For the maximum value, it was assumed that all elements with the [#] have a reinforcing effect, which was added to the value of all interdependencies with a reinforcing effect [e]. The average was then calculated to provide a clearer overview.

To define a main value, the minimum ("Sum with positive #") and maximum ("Sum with negative #") values of all influence factors were first calculated. The "Average main value" was then derived by calculating the average of the minimum and maximum values. This is illustrated in Figure 4.4.

In addition, a colour gradient from red to yellow to green has been integrated for better visualisation. The colour meanings correspond to the logic of the interdependency matrix. When looking at the total values of the right-hand column of the table, red means that the influence factor in the left-hand column has an unfavourable influence on many other influence factors (upper column) for the CE. Green means that the influence factor on the left-hand side tends to favour (strengthen) the other influence factors promoting the CE. Green therefore results when there are many reinforcing effects [e] in the row. Values in the middle are shown in yellow. These influence factors can generally have few effects, some reinforcing effects as well as some decreasing effects, or they can contain many elements with the defined interdependency "it can have a reinforcing or reducing effect" with the symbol [#]. Figure 4.4 shows a segment of the horizontal integration of the sums in the interdependency matrix.

Based on these classifications with colours, companies can recognise which influence factors are important for promoting the CE. The implementation of measures to strengthen these influence factors should be prioritised. As shown in Figure 4.4, the *Modularity of the product* is one of the influence factors that should be prioritised when defining measures, as it has a high numerical value in the right-hand column of the "Average main value". The same applies to the example of the influence factor *Circularity of the material flow*.

In the resulting interdependency matrix, the *Modularity of the product*, *R-strategy friendly joining techniques*, and *Reusability of components* have the highest horizontal average main values in the right-hand column. These influence factors therefore have the most positive effects on other influence factors. In contrast, the *Minimisation of air pollution*, *Minimisation of soil pollution*, and *Minimisation of material costs* have the lowest horizontal average main values (noted in the right-hand column). This means that these influence factors have a CE-fostering impact on only a few other influence factors or even a negative impact on other factors when considering CE.

When looking at the total values in the bottom row (see Figure 4.5), the significance is the opposite of the values in the right-hand column (see Figure 4.4). The

Category	Subcategory	Read from left to top	Reward system for take-back system	Explanatory examples	Margin per R-step (reference to costs)	Explanatory examples	Circularity of the materia	Explanato
Reprocessing	R-strategy reprocessing	Minimisation of reprocessing emissions	0	The idea o	#	The minim	%	The minim
		Minimisation of reprocessing costs	%	Lower pro	e	By minimi:	%	An excessi
		Positive influences [e]	1		10		14	
		negative influences [n]	0		0		2	
		Positive-or-negative influences [#]	0		4		1	
		Maximum: sum with all positive #	1		14		13	
		Minimum: sum with all negative #	1		6		11	
		Average main value	1		10		12	

Influence factors that are:

Positive by their nature (originally highlighted in *green*)

Negative by their nature and received a positive orientation through wording (originally highlighted in *red*)

Dividing lines between the areas of the matrix (overview of interdependencies and examples, and calculative values)

Highlighting in various shades of green (here in light grey) for high values of the numerical range calculated

Highlighting in red (here in dark grey) for low values of the numerical range calculated

FIGURE 4.5 Segment of the circular value creation influence factors interdependency matrix with vertical sums and average main values.

values in the bottom rows indicate whether an influence factor is strengthened or weakened by other influence factors. The total values in the bottom line are important for the identification of trade-offs. Figure 4.5 shows a segment of the vertical integration of the totals in the interdependency matrix.

Particularly low (red-shaded) overall values indicate the influence factors for which the strengthening or improvement of other factors can lead to a deterioration of the influence factor under consideration. The lower the total numerical value, the more likely it is that trade-offs will occur. Due to the logic of the matrix described above, negative numerical values are also possible. Negative numerical values are given a red colour. Influence factors with a low total value in the bottom row are a risk for the overall improvement towards a CE in a company.

For companies, this means that special consideration must be given to these influence factors. Where possible, measures to avoid a deterioration of these factors (e.g. influence factor A) and solutions for conflicting objectives should be defined and implemented. If it is not possible to resolve conflicting objectives, it must be evaluated whether the effects of improving the conflicting measures for the improvement of influence factors B-X outweigh the effects of the deterioration of influence factor A or whether the focus should be prioritised on avoiding a deterioration of influence factor A to ensure an improvement in the overall situation.

An example of a high vertical average main value in the bottom row of the developed interdependency matrix is the *Minimisation of air pollution*. A connection can be identified here. Since *Minimisation of air pollution* had one of the lowest horizontal average main values in the right-hand column, it is clear that this is an influence factor that is strongly affected by other factors but has little impact on other influence factors itself. In contrast, the *Minimisation of WIP costs* and the *Renewable raw material* have low vertical average main values in the bottom column. This implies that these influence factors are minimally influenced by other CE-promoting factors.

In theory, negative values would be possible. However, in this work, the calculated average main values of the interdependency matrix contained no negative values. In this case, the smallest value was zero and the largest was 22. The interdependency matrix includes the interdependencies: "It may or may not have an influence" [%] and "it can have a reinforcing or reducing effect" [#], which can vary as their names suggest. Due to these uncertainties, the average main values may change significantly when the matrix is applied to a specific case or company.

CONCLUSION AND FURTHER RESEARCH

The extension of the interdependency matrix strengthened the findings of Ruzicic et al. (2024).

As proven by Ruzicic et al. (2024) and this work, an interdependency matrix serves well to effectively depict the interdependencies between influence factors in circular value creation systems. This matrix is determined to be a valuable tool for mapping the correlations of these factors in the shift towards a circular value creation system (Ruzicic et al., 2024). The estimations, however, are contingent on the specific context and structure of a company, including its products, services, and production processes. In reference to Ruzicic et al., 2024, the interdependency matrix provides not only explicit estimations such as "has an effect" or "does not have an effect", but also includes "can, but does not necessarily have to have an effect".

Example-based explanations help the users of the matrix to understand the assessments of the interdependencies between the influence factors.

The interdependency matrix was refined and extended based on the findings. The findings will also be incorporated into a decision-support software tool developed as part of the research project Cyclometric, aimed at assisting engineers in the early stages of product development.

The result of the presented work on the development of an interdependency matrix does not claim to be exhaustive. The limitations stem from the literature reviewed and time constraints. Additionally, further research is required to validate the interdependency matrix and to incorporate more indicators and examples.

Ongoing adaptation of the interdependency matrix is necessary due to the continuous changes in framework conditions and global influences. To advance the analysis of quantification, further research is needed. By applying the matrix to a specific company, real data and values can be incorporated, facilitating its use in assessing planned and existing services, products, and their associated value creation systems. Companies can use the interdependency matrix as a supporting tool for transitioning to circular production and aligning their products and services with the principles of the CE (Ruzicic et al., 2024). By considering the interrelationships, companies can avoid potential negative consequences for the overall concept that may arise from focusing on supposed optimisations of a single aspect of a circular value creation system, while neglecting the broader perspective. While the previous interdependency matrix of Ruzicic et al. (2024) focused on presenting the interdependencies between influence factors of circular value creation systems, the extended interdependency matrix also includes a calculation logic for easier prisonisation and recognition of the influence factors with lots of dependencies. In this way, the extended interdependency matrix offers more benefits for application in industry.

ACKNOWLEDGEMENT

The authors would like to acknowledge the Federal Ministry of Education and Research (BMBF) sponsors the research project "Cyclometric" (02J21E030 ff.), which contributed to the results of this work.

REFERENCES

Abele, Eberhard. Chryssolouris, George. Sihn, Wilfried. Metternich, Joachim. ElMaraghy, Hoda. Seliger, Günther. Sivard, Gunilla. ElMaraghy, Waguih. Hummel, Vera. Tisch Michael and Seifermann, Stefan (2017). Learning factories for future oriented research and education in manufacturing, in: CIRP Annals, Volume 66. No. 2. pp. 803–826. DOI: 10.1016/j.cirp.2017.05.005

Aranda-Usón, Alfonso. Portillo-Tarragona, Pilar. Scarpellini, Sabina and Llena-Macarulla, Fernando. (2020). The progressive adoption of a circular economy by businesses for cleaner production: An approach from a regional study in Spain, *Journal of Cleaner Production*, Volume 247. No. 119648. pp.1–12. DOI: 10.1016/j.jclepro.2019.119648

Bag, Surajit. Yadav, Gunjan. Dhamija, Pavitra and Kataria, Krishan Kumar. (2021). Key resources for industry 4.0 adoption and its effect on sustainable production and circular economy: An empirical study, *Journal of Cleaner Production*, Volume 281. No. 125233. pp. 1–12. DOI: 10.1016/j.jclepro.2020.125233

Baldassarre, Brian. Keskin, Duygu. Diehl, Jan Carel. Bocken, Nancy. Calabretta, Giulia. (2020). Implementing sustainable design theory in business practice: A call to action, *Journal of Cleaner Production*, Volume 273. No. 123113. pp. 1–17. DOI: https://doi.org/10.1016/j.jclepro.2020.123113.

Elia, Valerio. Gnoni, Maria Grazia and Tornese, Fabiana. (2017). Measuring circular economy strategies through index methods: A critical analysis, *Journal of Cleaner Production*, Volume 142, Part 4. pp. 2741–2751. DOI: http://dx.doi.org/10.1016/j.jclepro.2016.10.196

European Commission. (2023) Industry 5.0: What this approach is focused on, how it will be achieved and how it is already being implemented. European Commission Website: https://research-and-innovation.ec.europa.eu/research-area/industrial-research-and-innovation/industry-50_en.

Ghisellini, Patrizia. Cialani, Catia and Ulgiati, Sergio. (2016). A review on circular economy: the expected transition to a balanced interplay of environmental and economic systems, *Journal of Cleaner Production*, Volume 114. pp. 11–32. DOI: https://doi.org/10.1016/j.jclepro.2015.09.007

Kadner, Suzanne. Kobus, Jörn. Hansen, Erik. Akinci, Seda. Elsner, Peter. Hagelüken, Christian. Jaeger-Erben, Melanie. Kick, Marlene. Kwade, Arno. Kühl, Carl. Müller-Kirschbaum, Thomas. Obeth, Dominik. Schweitzer, Katharina. Stuchtey, Martin. Vahle, Tilmann. Weber, Thomas. Wiedemann, Patrick. Wilts, Henning and von Wittken, Reinhard (2021). Circular Economy Initiative Deutschland (Ed.): Circular Economy Roadmap for Germany (Update December 2021). acatech/SYSTEMIQ, Munich/London, p. 15. DOI: https://doi.org/10.48669/ceid_2021-4

Kara, Sami. Hauschild, Michael. Sutherland, John and McAloone, Tim (2022). Closed-loop systems to circular economy: A pathway to environmental sustainability?, in: *CIRP Annals*, Volume 71. No. 2. pp. 505–528. DOI: 10.1016/j.cirp.2022.05.008

Kirchherr, Julian. Yang, Han-Hua Nadja. Schulze-Spüntrup, Frederik. Heerink, Maarten J. and Hartley, Kris. (2023). Conceptualizing the circular economy (revisited): An analysis of 221 definitions, *Resources, Conservation and Recycling*, Volume 194. No. 107001. pp. 1–12. DOI: 10.1016/j.resconrec.2023.107001

Koller, Jan. Grosse Erdmann, Julian. Herold, Maximilian and Döpper, Frank (2023). RemanLab – conceptualization and realization of a learning factory for remanufacturing, in: Proceedings of the 13th Conference on Learning Factories. Elsevier Inc., Reutlingen. DOI: http://dx.doi.org/10.2139/ssrn.4469183

Lee, Sir Yee Lee. Hu, Jiayao and K Lim, Ming (2024). Revisiting circular economy indicators: A circular supply chain perspective, *Journal of Purchasing and Supply Management*, Volume 30. No. 4. pp. 1–18 DOI: https://doi.org/10.1016/j.pursup.2024.100941

McDonough, William and Braungart, Michael (2002). *Cradle to Cradle. Remaking the way we make things.* NY North Point Press. ISBN: 9780865475878

Meidl, Rachel A. (2021). Disentangling Circular Economy, Sustainability and Waste Management Principles: James A. Baker III Institute for Public Policy of Rice University. pp. 1–6. DOI: 10.25613/884Q-M392

Nascimento, Daniel Luiz Mattos. Alencastro, Viviam. Quelhas, Osvaldo Luiz Gonçalves. Caiado, Rodrigo Goyannes Gusmão. Garza-Reyes, Jose Arturo. Rocha-Lona, Luis and Tortorella, Guilherme. (2019). Exploring Industry 4.0 technologies to enable circular economy practices in a manufacturing context, *Journal of Manufacturing Technology Management,* Volume 30. pp. 607–627. DOI: 10.1108/JMTM-03-2018-0071

OECD (2020), The Circular Economy in Cities and Regions: Synthesis Report, OECD Urban Studies, OECD Publishing, Paris. DOI: 10.1787/10ac6ae4-en

Potting, José. Hekkert, Marko. Worrell, Ernst and Hanemaaijer, Aldert (2017). Circular Economy: Measuring Innovation in the Product Chain (No. 2544). The Hague. Retrieved from PBL Netherlands Environmental Assessment Agency website: https://dspace.library.uu.nl/handle/1874/358310

Richardson, Katherine. Steffen, Will. Lucht, Wolfgang. Bendtsen, Jørgen. Cornell, Sarah E. Donges, Jonathan F. Drüke, Markus. Fertzer, Ingo. Bala, Govindasamy. Von Bloh, Werner. Feulner, Georg . Fiedler, Stehanie. Gerten, Dieter. Gleeson, Tom. Hofmann, Matthias. Huiskamp, Willem. Kummu, Matti. Mohan, Chinchu. Nogués-Bravo, David. Petri, Stefan. Porkka, Mina. Rahmstorf, Stefan. Schaphoff, Sibyll. Thonicke, Kirsten . Tobian, Arne. Virkki, Vili. Wang-Erlandsson, Lan. Weber, Lisa and Rockström, Johan (2023). Earth beyond six of nine planetary boundaries. *Science Advances,* Volume 9 No. 37. pp. 1–9. DOI: 10.1126/sciadv.adh2458

Rossi, Efigenia. Bertassini, Ana Carolina. Dos Santos Ferreira, Camila. Neves do Amaral, Weber Antonio and Ometto, Aldo Roberto (2020). Circular economy indicators for organizations considering sustainability and business models: Plastic, textile and electro-electronic cases, *Journal of Cleaner Production*, Volume 247. pp. 1–16. DOI: https://doi.org/10.1016/j.jclepro.2019.119137

Ruzicic, Nada. Braun, Anja and Ohlhausen, Peter (2024). Interdependency matrix to evaluate influence factors in circular value creation systems, in: Human Aspects of Advanced Manufacturing, Production Management and Process Control: Proceedings of the 15th International Conference on Applied Human Factors and Ergonomics and the Affiliated Conferences, Nice, France, 24–27 July 2024. Vol. 146. AHFE International, 2024. DOI: https://doi.org/10.54941/ahfe1005152

Saidani, Michael. Yannou, Bernard. Leroy, Yann and Cluzel, François (2017). How to assess product performance in the circular economy? Proposed requirements for the design of a circularity measurement framework, *Recycling,* Volume 2. No. 1:6. pp. 1–18. DOI: 10.3390/recycling2010006

Schmitt, Thomas. Wolf, Christopher. Lennerfors, Thomas Taro and Okwir, Simon. (2021). Beyond “Leanear” production: A multi-level approach for achieving circularity in a lean manufacturing context, *Journal of Cleaner Production*, Volume 318 No. 128531. pp. 1–13. DOI: 10.1016/j.jclepro.2021.128531

Schoonenboom, Judith and R. Burke, Johnson (2017). How to construct a mixed methods research design, Kölner Zeitschrift für Soziologie und Sozialpsychologie, Volume 69 (Suppl 2). pp. 107–131. DOI: 10.1007/s11577-017-0454-1

Suzanne, Elodie. Absi, Nabil and Borodin, Valeria. (2020) Towards circular economy in production planning: Challenges and opportunities, *European Journal of Operational Research*, Volume 287 No. 1. pp. 168–190. DOI: https://doi.org/10.1016/j.ejor.2020.04.043

Zucchella, Antonella and Previtali, Pietro (2019). Circular business models for sustainable development: A “waste is food” restorative ecosystem, *Business Strategy and the Environment*, Volume 28. pp. 274–284. DOI: https://doi.org/10.1002/bse.2216

5 Digitalization and High-End Manufacturing in Made in Italy Districts

Integrating Design Experiences in 5.0 Supply Chain

Gabriele Goretti and Benedetta Terenzi

MADE IN ITALY, IN BETWEEN INNOVATION AND CRAFTSMANSHIP PROCESSES

The term "Made in Italy" signifies an indication of origin that denotes products originating from Italy, as defined by EU regulations on non-preferential origin, particularly Italian Law No. 27, 2023, n. 206. This concept embodies a strong association with high quality, craftsmanship, and distinctive Italian design, acquiring significant cultural, symbolic, and commercial value internationally (Terenzi & Furin, 2020). Rooted in the Italian Renaissance, the evolution of "Made in Italy" reflects a synergy of creativity, craftsmanship, and innovation, ultimately evolving into a "metabrand" that transcends physical goods to evoke a broader idea of Italian style and culture recognized globally.

Recently, digitalization has become a critical factor in the evolution of this sector, essential for sustaining international competitiveness and influencing various production aspects. This importance is highlighted in the Confindustria report on digitalization and innovation in the Italian manufacturing sector and the European Commission's Joint Research Centre study on the impact of digital transformation on Italian manufacturing small- and medium-sized enterprises (SMEs).

Italy's furniture manufacturing districts consist of numerous independent SMEs, each specializing in specific tasks and typically interacting through subcontracting. A leading SME coordinates production, overseeing design, product engineering, and logistics. Since the late 1990s, the term "cluster," as defined by Porter (1998), has characterized these economic frameworks.

With the onset of the Lisbon Strategy in 2000, the European Commission transitioned toward a knowledge-based economy to enhance the competitiveness of European enterprises. Varaldo (2006) identifies key traits of this economy, including

DOI: 10.1201/9781003625360-5

the growing importance of knowledge and intangible assets within the global value chain (GVC), as well as the exchange of information and expertise. In light of the recent global financial crisis, this research explores organizational and operational models for a renewed "Made in Italy," integrating high craftsmanship with technological innovation.

Craftsmanship embodies the artistic skills and intangible values of production (Fuchs et al., 2015), while merging this expertise with contemporary innovation leads to what is termed "advanced craftsmanship" (Goretti, 2017). Some leading SMEs have initiated efforts to bridge the material dimension of their products with the immaterial realm of knowledge and skills, enhancing their identity and corporate heritage. In this context, digital archives play a crucial role in connecting traditional craftsmanship with contemporary creativity, inspiring new product designs, preserving evolving memories, and facilitating the application of historical knowledge in daily design activities, thereby becoming knowledge-centered (Lupo, 2019).

Although SMEs in the "Made in Italy" sector have become aware of the opportunities presented by digital transformation—such as efficiency gains, cost reduction, and increased flexibility—the Italian landscape remains heterogeneous, with a noticeable lack of digital skills. The digital transformation within manufacturing districts is characterized by common factors: technological innovation in production processes through technology transfer from other industries (e.g., automotive or fashion), safeguarding artisanal values while implementing technological advancements, and improvements in planning, production, and time-to-market. Additionally, the involvement of end users in communicating production values and customizing products is increasingly emphasized (Goretti, 2017).

DIGITALIZATION ACTION POINTS IN SME CLUSTERS: EU PROGRAMS AND ONGOING IMPLEMENTATIONS IN FURNITURE MANUFACTURING 5.0

Horizon 2020 is a significant program in which the European Union supports individual researchers, multinational cooperation projects, and SMEs by providing research and innovation funding. There are many projects in order to boost innovation and technological development in SMEs in the furniture sector, such as IN4WOOD, DITRAMA, and EQ-WOOD. Generally speaking, the furniture industry is a field where technology is not used extensively in production, and it is a low-tech industry (Otero-Neira et al., 2009). However, in order to increase brand value and strengthen competition, it is important to provide digital transformation within the company with technological developments.

Today, Manufacturing 5.0 represents the evolution of industrial manufacturing, moving beyond Industry 4.0. While the latter focuses on automation and the Internet of things (IoT), Manufacturing 5.0 emphasizes human–machine integration. This new approach aims to create a balance between human capabilities and advanced technologies, such as artificial intelligence (AI) and robotics to improve productivity, creativity, and sustainability (Lu & Shaharudin, 2024).

One of the main goals of Manufacturing 5.0 is to respect human beings by reintegrating creativity, intuition, and craftsmanship. In this context, advanced technologies support the worker without replacing him or her, making the manufacturing environment safer and more personalized. Digital technologies, such as AI, big data analytics (BDA), or Digital Twin (Terenzi & Vignati, 2020, help analyze complex data for more informed decisions, while numeric control machines and collaborative robotics (cobots) enable greater flexibility and precision. From a sustainable perspective, Manufacturing 5.0 works to reduce waste and optimize resource use, supporting the circular economy (Narkhede et al., 2024).

While the New European Bauhaus is a European Commission initiative launched to combine sustainability with aesthetics and inclusiveness, drawing inspiration from the principles of the 20th-century Bauhaus movement. It aims to create a bridge between digitization and creativity, promoting projects and spaces that are both innovative, ecological, and socially inclusive. This initiative encourages collaboration among artists, architects, designers, engineers, and scientists to develop sustainable solutions that address environmental and social challenges.

The European Green Deal, a sustainable growth strategy of the European Commission, designed as a baseline to guide policies, investments, and initiatives aimed at achieving an inclusive, economically sustainable, and socially equitable green transition, includes a number of key significant goals, such as reducing greenhouse gas emissions, protecting biodiversity, promoting the circular economy, enhancing energy efficiency, and transitioning to renewable energy sources. The Green Deal strategy goes beyond environmental issues, integrating aspects of social justice, and technological innovation to ensure that all economic sectors, including industry, agriculture, and transportation, contribute to sustainable growth (Oberthür & von Homeyer, 2022).

In fact, it is certainly important to emphasize the key role of the dual digital and green transition. In this sense, SMEs and startups in different sectors can drive substantial change through eco-innovation, addressing challenges such as climate change and resource efficiency. Digital tools can play an important role in supporting SMEs in achieving their green goals. Digitization can promote resource efficiency, improve transparency of green practices, and provide an effective channel for training entrepreneurs. SMEs can accelerate their transition to sustainable models by implementing eco-innovations based on digital technology, such as carbon footprint verification technologies, tracking sustainable materials in supply chains through distributed ledger technology, or using green sensors connected to IoT devices to promote green practices also through physical and digital channels.

SME DIGITALIZATION, FROM PRODUCT TO PHYGITAL INTERACTION

Digitalization of SMEs is crucial in building inclusive and resilient economies and societies. Ensuring the uptake of digital technologies by all SMEs and entrepreneurs is central to fully unlocking the potential of the digital revolution at large. SMEs are the backbone of most OECD economies (OECD, 2019). Across the OECD, they account for 99% of all businesses and between 50% and 60% of value added. SMEs are an important source of employment for communities, with almost one person out of three being employed in a microfirm with less than ten employees and two out of three in an SME. SMEs are also strategic actors in GVCs, often playing a key role in large firms' supply chains (OECD, 2019).

In the last years, digital transformation has revolutionized the way companies conduct business, create relationships with consumers, suppliers, and other stakeholders (Bresciani et al., 2021 Scuotto et al., 2019), and foster business model innovation and customer value creation. We can succinctly define digital transformation in SMEs as the way a company employs digital technologies to develop a new digital business model in order to create more value for the company (Verhoef et al., 2021). Such transformation affects business processes, operational routines, and organizational capabilities (Li, Su, Zhang, & Mao, 2018).

Digitalization is shaping new ways of interaction between consumers and businesses (Taiminen & Karjaluoto, 2015). In particular, consumers have access to dozens of media channels through which to communicate actively and effortlessly with firms and other consumers, passing through a rapidly increasing number of touchpoints on their customer journey (Terenzi & Vignati, 2021, Verhoef et al., 2021).

The OECD "Digital for SMEs" (D4SME) Global Initiative aims to promote knowledge sharing and learning on how to enable businesses of all sizes to benefit from the digital transformation. To this aim, the initiative encourages the sharing of experiences and good practices to support the digital transition of all SMEs, according to their needs and level of digital maturity and ambitions. The initiative thus targets SMEs that are at the beginning of their digitalization journey, usually starting from the uptake of basic digital tools (e.g., having a functional website, managing effectively social media profiles), as well as firms that have already embedded some or several digital technologies in their processes and are looking for the "next steps" to take (e.g., digitalization of core business functions, uptake of frontier technologies such as AI, blockchain, augmented reality (AR), virtual reality (VR)).

The ability to perceive new digital opportunities, to change customers' interactions, and to co-create value with them implies changes in existing routines or resource configurations as well as building new capabilities. Filling these gaps is important, because digital technologies are important competitive tools for SMEs, and neglecting them is risky and may seriously compromise firm survival. Digital technologies provide support for value creation and customer engagement, which are critical success factors for SMEs. However, the use of technologies is not without difficulties and requires a change management and vision that have foundations in the acquisition of new organizational and marketing capabilities (Matarazzo, et al., 2021).

The label "phygital" (physical plus digital) has been proposed to describe how companies try to engage customers due to the exploitation of smart technologies. The research into phygital's theoretical development in SMEs is lacking. The in-depth analysis allows for the identification of four main aspects: (1) objects and applications, (2) context (space/place), (3) customer journey, and (4) shopping (retail) experience. Using the antecedents, decisions, and outcomes model, we offer an interpretation of the phenomenon in a more comprehensive framework. The development and use of physical and digital objects and applications (i.e., phygital resources) in a phygital context (i.e., spaces and places) affect the customer journey in a phygital manner to enable the development of new forms of phygital experience. We suggest specific research avenues to develop the phygital construct (Mele, Spena, Marzullo, et al., 2023).

A new label - phygital (physical plus digital) - has been proposed to describe how companies try to engage customers through a combination of the virtual and real (Ballina et al., 2019; Clauzel et al., 2019) due to the exploitation of smart technologies. The concept of "phygitality" as a new form of "hybrid consumption experience" has also been proposed (Klaus, 2021, p. 1). Immersive technologies, such as AR and VR, tech solutions based on AI, Internet of things (IoT), Internet of everything (IoE), conversational agents (chatbot), and customer relationship management solutions, are just some of the digital applications that enable this transformation (Mele & Russo Spena, 2022). However, the term phygital lacks a clear definition and offers several interpretations. This conceptual vagueness complicates academic discourse on the opportunities and limits of the phygital phenomenon and its transformation (Mele, Spena, Marzullo, et al., 2023).

Moreover, the notion of playful interaction emerges as a potent means of knowledge acquisition and entertainment. By playful interaction, we mean an activity in which an effort is required by a user that is rewarded with a pleasant improvement of skills. The process is also characterized by the presence of stimulated emotions through and thanks to the use of the product (physical or digital) (Forlizzi & Battarbee, 2004).

MANUFACTURING 5.0 AND PSS STRATEGIES IN SMES

SMEs integrating Industry 5.0 from both technological and organizational perspectives aim to enhance traditional craft-oriented production methods by making them more digital and efficient. This strategic approach allows these companies to adopt flexible business models and undergo "smart" transformations in logistics, exemplified by initiatives such as Smart Lifecycle, Smart Supply Chain, and Smart Factories. Industry 5.0 signifies the convergence of production, information technologies, and the Internet, promoting the integration of Internet services that enhance company competitiveness (Rauch et al., 2020). This integration facilitates the emergence of profitable product service systems (PSS), which represent a dual focus on service and production-oriented thinking. According to OECD research 2021 (OECD, 2021a; OECD, 2021b) , such simultaneous development leads to improved profitability for companies (Mueller & Thoring, 2012). In today's competitive landscape, businesses lacking technological integration face significant challenges; merely relying on fast, high-quality, and low-cost production is no longer sufficient. Innovative and digital production strategies have become increasingly critical for success (Rauch et al., 2020).

Numerous Italian furniture SMEs exemplify effective implementation of contemporary technologies across various production sectors. However, traditional craftsmanship and human factors remain central to the identity of these design-oriented SMEs, highlighting the unique blend of heritage and innovation that characterizes Italian manufacturing (Goretti et al., 2019).

Digital transformation has far-reaching implications for archives, which have gained importance across multiple research domains. They serve as repositories of organized memory, facilitating the dissemination of knowledge across various disciplinary, professional, and technological spaces (Gilliland, Lau, & McKemmish, 2021). The analysis reveals common experimental territories in the role of digital archives as catalysts for the introduction of additional tools integral to digital transition, in conjunction with daily production and communication technologies (Giannachi, 2016). Furthermore, these archives inspire design choices aligned with the technical specifications employed in production, allowing SMEs to develop pathways toward advanced craftsmanship (Fry et al., 2017; Goretti, 2017). This evolution fosters new processes of design-driven innovation in both business-to-business (B2B) and business-to-consumer (B2C) contexts.

The wealth of information collected regarding products and their digital visualizations in these digital archives, along with the interactive services offered to end users, can be incorporated into specific product lifecycle management (PLM) processes. PLM serves as a product management model capable of defining the phases and timelines of production while digitalizing the tangible and intangible know-how of company workers (Brozzi et al., 2018). The focus on time-to-market remains one of the primary interests for SMEs. By developing interactive systems that engage and inform end users, these production systems enable companies to maintain the integrity of their production timelines and processes.

Lago Design Case Study, Manufacturing 5.0, and PSS Strategy

Lago Design, a brand with a rich heritage that dates back to the late 19th century, initially focused on interior design for villas and churches. Today, it has successfully expanded into the international market, now in its fourth generation. The company has embraced technological innovation by implementing a 3D Configurator, which empowers customers to actively participate in the production design process. This service-oriented production model prioritizes customer preferences, delivering tailored solutions that enhance both product and service dimensions. Ultimately, the PSS enriches the customer experience and strengthens brand value.

Lago perceives design as an essential instrument for understanding and addressing the complexities of contemporary life. By engaging customers in the production process, the company remains attuned to current market trends while simultaneously enhancing traditional craftsmanship through innovative approaches. This adaptability in production and service offerings exemplifies an effective brand strategy that resonates with evolving consumer needs.

The COVID-19 pandemic prompted significant changes in how individuals utilized their homes, with many transforming living spaces into home offices. As a result, there was a marked increase in demand for furniture that could effectively delineate workspaces from living areas. Research indicates that during this period, the import rate of upholstered furniture consumption surged to 34% (The World Upholstered, 2019). SMEs are particularly susceptible to fluctuations in their environment, making resilience a critical factor for their continuity during the pandemic (Zuperkiene et al.,

2021). However, SMEs also have the unique advantage of rapidly adapting to these changes and emerging stronger by learning from adversity. The introduction of PSS can significantly enhance operational efficiency and management capabilities, allowing SMEs to respond effectively to market demands.

Lago's deeply ingrained digital business culture, reflective of its historical roots, enabled the company to meet customer demands for furniture renovation during the pandemic, despite maintaining 500 physical stores. The brand developed an omnichannel system termed "shopping experience," which integrates design and customer needs through 3D visualizations, real-time store visits, and access to the latest collections (Archiproducts, 2020). In response to shifting consumer preferences, Lago swiftly launched its "Home Office" collection and expanded its furniture renovation options. Additionally, the company offers online consultations between customers and designers, flexible product design services, and mobile applications to facilitate the design process.

According to company communications, Lago demonstrated remarkable resilience by opening a new branch in Shanghai during the pandemic. The Digital SME Alliance highlights that the crisis has accelerated opportunities for digitalization and outlines key areas for SMEs to focus on developing digital skills, implementing change programs to mitigate unemployment risks, investing in digitalization, facilitating cross-border purchasing, and securing support for digital transformation (European Digital SME Alliance, 2020).

Throughout the pandemic, business operations, meetings, shopping, and marketing increasingly shifted to virtual environments. Companies expanded their focus on technology-driven digitalization, particularly within service areas, transforming product presentations and marketing strategies. Digital catalogs and mobile applications enable customers to gain a deeper understanding of product value and easily access company offerings. Furthermore, advancements in one area of a company can have positive ripple effects in other domains, fostering integrated management innovation.

The pervasive digital approach of Lago is particularly emphasized by its 3D configurator (https://configurator.lago.it), which allows users to explore products in detail and customize them through modular systems. This configurator is based on a parametric digital archive that connects the services offered and user choices to an enterprise PLM system, ensuring optimized order integration within the manufacturing process. By streamlining production phases and timelines, Lago effectively enhances both the customer experience and operational efficiency.

TOWARD IMMERSIVE VIRTUAL EXHIBITIONS IN FURNITURE AND INTERIOR DESIGN

The concept of augmented exhibitions, as explored in this research, represents a transformative approach to the presentation of artifacts and information. These exhibitions utilize digital media - such as images, videos, or interactive 3D models -integrated within real spaces, thereby creating a hybrid environment that engages audiences in unprecedented ways. This evolution has gained momentum, particularly with the proliferation of smartphones and the advent of all-in-one wearable VR visors, which have enabled more exhibitions to transform seemingly empty showcases into dynamic platforms for digital interaction.

Unlike traditional digital media forms, such as LED screens or projection displays, augmented exhibitions are characterized by their ability to seamlessly combine real and virtual objects within a shared environment. This interaction is not only real-time but also responsive, allowing for the alignment of physical and digital elements. As exhibition spaces are inherently visual, they provide an ideal context for the application of AR, a technology that emphasizes visual engagement and interaction.

The notion of the virtual experience further elaborates on the impact of augmented exhibitions. This experience encompasses the user reactions and behaviors as the reality of a virtual object supersedes the physical space around it. Essentially, the augmented experience can be conceptualized as a layering of information over physical environments, enhancing the user perception and interaction with their surroundings.

According to Manovich (2006, p. 13):

> Augmented space research gives us new terms with which to think about previous spatial practices. Before we would think of an architect, a fresco painter, or a display designer working to combine architecture and images, or architecture and text, or to incorporate different symbolic systems in one spatial construction, we can now say that all of them were working on the problem of augmented space. The problem, that is, of how to overlay physical space with layers of data.

The system of data generated from immersive experiences and user interactions with exhibited digital products and surrounding spaces remains a topic of ongoing debate and experimentation. Understanding how both quantitative and qualitative information can be effectively integrated into production systems is essential for enhancing efficiency and productivity.

Focusing more closely on the technologies employed, we can distinguish between AR and VR. AR serves as a bridge between virtual exhibits and real-world experiences, utilizing technology to analyze physical spaces and facilitate appropriate interactions with virtual objects. In contrast, VR immerses users in a fully digital realm, where all perceived objects are digitally created. This immersive experience, while engaging, is detached from any specific physical location, representing an extreme iteration of AR (Shilong, 2021).

The ongoing digitalization processes within the furniture production supply chain present fertile ground for experimenting with these virtual experiences. Such digital transformations should not be viewed solely as means of production efficiency; rather, they represent opportunities for developing user-centered promotional and informative experiences.

This research presents two case studies focusing on the application of AR and VR in the context of manufacturing brands. The aim is to develop innovative services and immersive experiences tailored for both professional audiences (B2B) and end users (B2C).

Colux Case Study, Immersive Technologies for Co-design Strategies

The Colux Project represents a pioneering collaborative initiative involving three notable furniture manufacturers - Marioni and Savio Firmino - alongside a real estate firm and two prominent universities, the University of Florence and the University of Siena in Tuscany, Italy. Supported by the Italian National Research Consortium (CNR), the project has garnered funding through the POR FESR 2014 - 2020 program of the Tuscany government, as part of the broader European Horizon 2020 framework.

Central to the Colux Project is the application of interactive technologies, specifically VR and AR, aimed at developing an interactive platform dedicated to co-designing products and physical environments within the furniture and home accessories sector. This innovative platform facilitates a dynamic workspace that enables the rapid and coordinated sharing of information among stakeholders across the production chain, including furniture manufacturers, architectural firms, design entities, and technical installers. By positioning furniture manufacturers at the core of this collaborative ecosystem, Colux enhances synergies among various actors involved in the design and production processes.

The primary goal of the Colux Project is to bolster the growth of high-end manufacturing companies within the furniture and interior design sector, thereby increasing their competitiveness in international markets. The project was officially launched at the Salone del Mobile in Milan in 2023, marking a pivotal moment in its introduction to the market. Through this initiative, Colux seeks to redefine the landscape of furniture design and production, fostering innovation and collaboration among industry participants.

Incorporating VR technology into the design process offers significant advantages, particularly in enhancing users' spatial perception and facilitating spatial design activities. VR environments provide users with richer visual information compared to traditional desktop settings, though this can also lead to challenges in processing that information efficiently. As indicated by Hwan and Kyung (2022), there is an ongoing investigation into whether VR distinctly influences the spatial design experience when compared to more conventional design environments.

Moreover, the potential of information and communication technologies (ICT) within the interior design field is substantial, particularly through the development of software and hardware tools for VR. Utilizing these ICT tools allows designers to achieve a more realistic perception of their proposals right from the conceptual stage. Enhancements in ICT software tools designed for VR can lead to increasingly realistic virtual environments, thereby improving the overall design experience.

The Colux Project exemplifies this innovative approach to interior design by employing cutting-edge ICT technologies, as detailed by Kaleja and Kozlovaska (2017). By integrating the latest advancements in VR and AR, Colux not only enriches the design process but also fosters a collaborative environment that bridges the gap between conceptualization and realization. This holistic approach not only enhances the creativity and efficiency of designers but also aligns with the evolving demands of the high-end furniture market.

The Colux Project stands as a significant advancement in the integration of VR and AR technologies within the furniture and interior design sector. By promoting collaborative co-design processes and utilizing sophisticated ICT tools, Colux aims to elevate the competitive landscape for high-end manufacturers. Through this initiative, the project is poised to foster a new era of innovation and collaboration in furniture design, ultimately reshaping the industry's future.

VR Exhibition for Savio Firmino at Salone del Mobile Shanghai

Savio Firmino, within the inaugural edition of Salone del Mobile in Shanghai in 2017, presented an accessible VR experience for retail customers, distributors, architects, and interior designers. This straightforward yet functional kit, offered at a cost-effective price, comprises a VR viewer and a smartphone, immersing users in a virtual environment with limitless customization options.

Designed to be portable and user-friendly, the 3D-box VR kit features a quick setup process and is supported by a team of technological experts, enabling users to visualize any interior project in VR. The accompanying application allows users to navigate freely within the space, modify furniture colors, and adjust materials and finishes.

By integrating tools of VR with AR, Savio Firmino aimed to assist its retail customers in assessing the impact of their projects in relation to both the furniture and the surrounding environment. Retail clients can provide their customers with the opportunity to preview and evaluate individual pieces of furniture or complete interior designs from the comfort of their homes, ultimately facilitating the selection and purchase process.

"We aim to make virtual reality an accessible tool for our team, empowering us to showcase a complete home furnishing experience with our products to both designers

FIGURE 5.1 3D-box virtual reality for Savio Firmino. (Savio Firmino, 2017.)

and private customers," stated CEO Cosimo Savio. "In the retail sector, our 3D-Box enhances the customer experience by combining artisanal craftsmanship with cutting-edge technology. This unique integration positions us as the first company worldwide to reach such an advanced level of virtual reality in furniture design" (seminar at DESTEC Department, 7 June 2024) (Figure 5.1).

DISCUSSION AND CONCLUSIONS

The digitalization of SMEs stands as a centric topic in building inclusive and resilient economies and societies. Although there has been a steady upward trend in the uptake of digital technologies by SMEs over the last decade, the gaps with large businesses remain large, contributing to inequalities among firms and, in turn, people and places. At the firm level, digital gaps are strongly associated with gaps in productivity, scaling up, innovation, and growth, all of which impact not only aggregate economic growth, but also inclusion and societal well-being.

SMEs face several barriers in their journey to digitalization. Lack of information and awareness, digital skills gaps, and insufficient capital to finance the transformation represent long-standing hurdles to going digital for a large share of the SME population. In addition, smaller firms face heightened difficulties in overcoming challenges in the broader business environment, such as rapidly changing regulatory frameworks, dealing with digital security and privacy issues, or simply accessing high-quality and affordable digital infrastructure. Smaller firms are also less aware of public programs that they could leverage at low or no cost to sustain their digital transformation.

In the Italian context, the development and deployment of digital archives focus primarily on enhancing manufacturing efficiency, quality certification, and the preservation of traditional know-how. Many examples as Lago or Savio Firmino exemplify a significant advancement in this arena, employing product-based PSS or VR and AR to create inclusive visualization and new user experiences. Although this approach has seen positive experimentation at relevant trade fairs since 2019, the broader application of user experiences based on digital archive systems remains sporadic within the Italian framework, often lacking well-defined objectives.

The experimentation highlighted in these cases offers valuable insights into the relationship between real and virtual experiences. The concept of a "phygital" environment - a blend of physical and digital experiences - serves as a critical area of exploration. This interplay allows for an enhanced understanding of products and spaces, bridging the gap between tangible and intangible dimensions of design.

By developing services that actively engage users, manufacturers can enhance visibility and showcase the quality of their craftsmanship through new communicative languages. Such an approach not only enriches the consumer experience but also fosters a deeper connection between the user and the product, promoting a sense of engagement that transcends traditional boundaries.

Reaching new audiences and expanding into untapped market segments are crucial for high-end manufacturing systems. This expansion allows manufacturers, particularly those rooted in artisanal traditions, to integrate more fully into the GVC, thereby facilitating growth and sustainability. Importantly, technology should be viewed as a supporter of craftsmanship rather than a replacement. By leveraging technological advancements, artisans can enhance their creative processes while preserving the intrinsic emotional values associated with their products.

REFERENCES

Archiproducts. (2020, November 13). Lago Shopping Experience. www.archiproducts.com/en/news/lago-shopping-experience_79600

Ballina, F.J., Valdes, L., and Del Valle, E. (2019). *The phygital experience in the smart tourism destination. International Journal of Tourism Cities*, 5(4), pp. 656–671.

Bresciani, S., Huarng, K.H., Malhotra, A., and Ferraris, A. (2021). Digital transformation as a springboard for product, process and business model innovation. *Journal of Business Research*, 128, pp. 204–210. https://doi.org/10.1016/j.jbusres.2021.02.003

Brozzi, R., D'Amico, R.D., Pasetti Monizza, G., Marcher, C., Riedl, M., and Matt, D. (2018). Design of self-assessment tools to measure industry 4.0 readiness – A methodological approach for craftsmanship SMEs. In Chiabert, P., Bouras, A., Noël, F. and Ríos, J. (eds), *Product Lifecycle Management to Support Industry 4.0*, PLM 2018 IFIP Advances in Information and Communication Technology, Springer, Cham, pp. 566–578. https://doi.org/10.1007/978-3-030-01614-2_52

Clauzel, A., Delacour, H., and Liarte, S. (2019). When cuisine becomes less Haute: The impact of expert ratings on consumers' legitimacy judgments. *Journal of Business Research*, 105, pp. 395–404.

European Digital SME Alliance. (2020, May). COVID-19 Economic Recovery: A Recovery Strategy Building on Digital.

Forlizzi, J., and Battarbee, K. (2004). *Understanding Experience in Interactive Systems.* DIS Proceedings of the 5th conference on Designing interactive systems: processes, practices, methods, and techniques, pp. 261–268. ACM, New York. https://doi.org/10.1145/1013115.1013152

Fry, A., Goretti, G., Ladhib, S., Cianfanelli, E., and Overby, C. (2017). Advanced Craft, integrated with the Saper Fare – The role of Intangible value, and the centrality of the artisan in high-quality 21st century artisanship. *Cuadernos del Centro de Estudios en Diseño y Comunicación*, 18(64), pp. 255–276. https://dialnet. unirioja.es/servlet/articulo?codigo=7325457.

Fuchs, C., Schreier, M., and Van Osselaer, S.M. (2015). *The handmade effect: What's love got to do with it? Journal of Marketing*, 79(2), pp. 98–110.

Giannachi, G. (2016), *Archive everything – Mapping the everyday*, MIT Press, Cambridge (MA).

Gilliland, A.J., Lau, A.J., and McKemmish, S. (2021). Pluralizing the archive. In Fujiyoshi, K. (ed.), *Archives for Maintaining Community and Society in the Digital Age*, Springer, Singapore, pp. 61–70. https://doi.org/10.1007/978-981-15-8514-2_7

Goretti, G. (2017). *Advanced craftsmanship – Maestria avanzata – Percorsi di progetto tra innovazione e tradizione artigianale nei sistemi manifatturieri toscani.* Aracne, Roma.

Goretti, G., Cianfanelli, E., Terenzi, B., and Tufarelli, M. (2019). Advanced Craftsmanship nell'arredo. Il comparto toscano tra tradizione e innovazione 4.0 / Advanced Craftsmanship in furniture. The Tuscan sector between Tradition and Innovation 4.0. *DIID Journal, Design Actually*, 62/63, pp. 26–33.

Hwan, K., and Kyung Hoon, H. (2022). *Understanding Design Experience in Virtual Reality for Interior Design Process.* POST-CARBON – Proceedings of the 27th CAADRIA Conference, Sydney, 9—15 April 2022, pp. 59–68.

Kaleja, P., and Kozlovská, M. (2017). *Virtual Reality as Innovative Approach to the Interior Designing. Selected Scientific Papers – Journal of Civil Engineering*, 12(1), pp. 109–116.

Klaus, P.P. (2021). *Phygital–the emperor's new clothes? Journal of Strategic Marketing*, 32(1), 1–8.

Li, L., Su, F., Zhang, W., and Mao, J. (2018). Digital transformation by SME entrepreneurs: A capability perspective. *Information Systems Journal*, 28, pp. 1129–1157.

Lu, H., and Shaharudin, M.S. (2024). Role of digital transformation for sustainable competitive advantage of SMEs: a systematic literature review. *Cogent Business & Management*, *11*(1), 1–16. https://doi.org/10.1080/23311975.2024.2419489

Lupo, E. (2019). Design e Cultural driven innovation, in *i+ Diseño | Revista científico-académica internacional de innovación, investigación y desarrollo en Diseño*, 14, pp. 120–132. https://doi.org/10.24310/ Idiseno.2019.v14i0.7085

Manovich, L. (2006). The poetics of augmented space. *Visual Communication,* 5(2), pp. 219–240.

Matarazzo, M., Penco, L., Profumo, G., and Quaglia, R., (2021) Digital transformation and customer value creation in Made in Italy SMEs: A dynamic capabilities perspective. *Journal of Business Research*, 123, pp. 642–656.

Mele, C., and Russo Spena, T. (2022). The architecture of the phygital customer journey: a dynamic interplay between systems of insights and systems of engagement. *European Journal of Marketing*. 56(1), pp. 72–91. https://doi.org 10.1108/EJM-04-2019-0308

Mele, C., Spena, T.R., and Marzullo, M. (2023). The phygital transformation: a systematic review and a research agenda. *Italian Journal of Marketing*, 2023, pp. 323–349. https://doi.org/10.1007/s43039-023-00070-7

Mueller, R.M., & Thoring, K. (2012). *Design Thinking Vs Lean Startup: A Comparison of Two User-driven Innovation Strategies*. Proceedings of 2012 International Design Management Research Conference. https://doi.org/10.13140/2.1.1834.4647

Narkhede, G., Chinchanikar, S., Narkhede, R., and Chaudhari, T. (2024). Role of Industry 5.0 for driving sustainability in the manufacturing sector: an emerging research agenda. *Journal of Strategy and Management*. www.emerald.com/insight/1755-425X.htm https://doi.org/10.1108/JSMA-06-2023-0144

Oberthür, S., and von Homeyer, I. (2022). From emissions trading to the European Green Deal: the evolution of the climate policy mix and climate policy integration in the EU. *Journal of European Public Policy*, *30*(3), pp. 445–468. https://doi.org/10.1080/13501763.2022.2120528

OECD (2019), OECD SME and Entrepreneurship Outlook 2019: Policy Highlights, www.oecd.org/industry/smes/SME-Outlook-Highlights-FINAL.pdf.

OECD (2021a), OECD SME and Entrepreneurship Outlook 2021,OECD Publishing, Paris

OECD (2021b), The Digital Transformation of SMEs, OECD Studies on SMEs and Entrepreneurship, OECD Publishing, Paris https://dx.doi.org/10.1787/bdb9256a-en.

Otero-Neira, C., Lindman, M.T., and Fernández, M.J. (2009). Innovation and performance in SME furniture industries: An international comparative case study. *Marketing Intelligence and Planning*, 27(2), 216–232. https://doi.org/10.1108/02634500910944995

Porter, M.E. (1998). *Clusters and the New Economics of Competition*. Harvard Business Reviewb Boston (USA)

Rauch, E., Vickery, A.R., Brown, C.A., and Matt, D.T. (2020). SME requirements and guidelines for the design of smart and highly adaptable manufacturing systems. *Industry 4.0 for SMEs: Challenges, Opportunities and Requirements*. Cham: Springer. https://doi.org/10.1007/978-3-030-25425-4_2

Scuotto, V., Serravalle, F., Murray, A., and Viassone, M. (2019). The shift towards a digital business model: A strategic decision for the female entrepreneur. In *Women Entrepreneurs and Strategic Decision Making in the Global Economy*. IGI Publishing, New York (USA), pp.24.

Shilong, T. (2021). *Virtual Experience in Augmented Exhibition. Virtual technology application in real museums and its impact on exhibition spaces*. Politecnico di Milano. www.politesi.polimi.it/handle/10589/171010

Taiminen, H.M., and Karjaluoto, H. (2015). The usage of digital marketing channels in SMEs. *Journal of Small Business and Enterprise Development*, 22(4), pp. 633–651. https://doi.org/10.1108/JSBED-05-2013-0073

Terenzi, B., and Furin, E., (2020). Dal design Made in Italy al design Made in ... Umbria. L'identità territoriale per l'innovazione del settore ceramico. *MD Journal*, 6, pp. 74–85.

Terenzi, B., and Vignati A. (2020). *Digital Transformation in Product Service System for Kids. Design Tools for Emerging Needs. Integrating People and Intelligent Systems*. Advances in Intelligent Systems and Computing, 1131, pp. 228–234, Springer, Cham. https://doi: 10.1007/978-3-030-39512-4_36

Terenzi, B., and Vignati, A. (2021). Smart retail and phygital customer journey in the kids and toys stores. In: Russo, D., Ahram, T., Karwowski, W., Di Bucchianico, G. and Taiar, R., *Intelligent Human Systems Integration 2021*. 1322, pp. 709–714, Springer, Cham. https://doi: 10.1007/978-3-030-68017-6_105

The World Upholstered Furniture Industry 2010-2019 and COVID-19's Impact During 2020-2021. Business Wire.

University of Pisa, Savio Firmino Seminar at Phd program of DESTEC Department at University of Pisa, 7 June 2024. https://destec.unipi.it/sites/default/files/2024-05/locandina_7_14_06_2024_eng_logo_light.pdf

Varaldo, R. (2006). Il nuovo modello competitivo e aziendale dei distretti industriali. *Economia e Politica Industriale*, 1, Milano: FrancoAngeli.

Verhoef, P.C., Broekhuizen, T.L., Bart, Y., Bhattacharya, A., Qi Dong, J., Fabian, N.E., and Haenlein, M. (2021). Digital transformation: A multidisciplinary reflection and research agenda. *Journal of Business Research*, 122, 889–901.

Zuperkiene, E., Simanskiene, L., Labanauskaite, D., Melnikova, J., and Davidaviciene, V. (2021). The COVID-19 Pandemic and Resilience of SME's in Lithuania. *Entrepreneurship and Sustainability Issues*, 8(3), 53–65. https://doi.org/10.9770/jesi.2021.8.3(4)

6 Investigating the Convergence of Digital Twins, the Metaverse, and Sustainability

A Study and Framework for Smart Cities

Mascha Samkova, Panagiotis Markopoulos, Georgios Markopoulos, and Akash Nandi

INTRODUCTION

Since the 1990s, advancements in internet technologies have significantly improved speed, stability, and bandwidth, catalyzing a revival of once-neglected technologies such as artificial intelligence (AI), business analytics, and decision support systems. Previously hindered by limited data volumes, these technologies are now thriving alongside the emergence of novel innovations like virtual reality and augmented reality (VR/AR), digital twins, metaverse, and building information modeling (BIM). In a powerful combination, these advancements foster optimism about past limitations and envision a future where technology can build a sustainable, responsible, and rewarding world for all (Markopoulos and Luimula, 2020).

Revived and emerging technologies' fast adaptation has led to transformative applications in the world economy. Financial innovation, including cryptocurrencies, autonomous transportation systems, digital twins for tracking and safety, expert systems in health care, and wearable technology in the fashion industry, all exemplify the diverse utilities of revived and emerging technologies. Other sectors of the economy, like education, entertainment, and architecture, have already benefited from virtual worlds. In only 10 years, these virtual innovations have evolved into broader ecosystems, producing smart buildings, vehicles, roads, and ultimately, smart cities, which collectively reshape the physical built environment today and tomorrow.

DOI: 10.1201/9781003625360-6

THE PROMISE OF SMART CITIES IN THE METAVERSE

In urban development, intelligent cities represent a vision bringing in harmony advanced technology with sustainability and social responsibility. While none of these cities have yet been fully realized, the methodology holds transformative potential. The metaverse, a shared virtual world, opens the world of rewarding hands-on prototypes for smart city development. Developers can analyze and refine human behavior, performance, and well-being within digital environments before implementing solutions in the physical world (Luimula, 2021).

In the following, carefully selected cases of metaverse-based urban projects are discussed.

Case 1. Liberland: The Cyber–Urban Experiment

One of the most fascinating examples of a metaverse smart city for its diplomatic objective is Liberland, an initiative spearheaded by private investors. Liberland exists as a unique response to geopolitical complexities, utilizing an unclaimed and disputed tract of land situated between Croatia and Serbia. This area of 7.2 square kilometers, being of limited use to either country due to its contested status, became the foundation for a virtual city, respecting the two countries' territorial integrity. Declared a "virtual nation", the Free Republic of Liberland was established on 13th April 2015. Today, one decade later, it remains unrecognized by international organizations or governments but formally claims sovereignty over the small area of land (Wadham, 2015; Haristianti and Murdowo, 2023).

The virtual nation Liberland distinguishes itself by its unconventional governance model. It has no official government, taxes, public services, or physical residents. Instead, it thrives as a digital society, attracting libertarian-minded individuals eager to explore new ways of living and working. Investors and developers capitalize on these ideals by creating a thriving virtual environment where citizenship is experienced digitally in the common language, English, as lingua franca.

To bring this vision to life, the British architecture firm *Zaha Hadid Architects*, named after the Iraqi-born British architect Zaha Hadid DBE RA, designed a "cyber–urban" cityscape that embodies futuristic aesthetics and cutting-edge design principles, as shown in Figure 6.1. Liberland exists entirely within the metaverse, featuring meticulously designed spaces and environments that skillfully blur the line between reality and virtuality (Schumacher, 2022).

Architectural Innovations in Liberland

Zaha Hadid Architects' vision for Liberland integrates striking visual and functional elements, offering a glimpse into the possibilities of digital urbanism. The panoramic design of Liberland showcases a harmonious blend of organic forms and avant-garde structures (Finney, 2022). Citizens, represented as avatars, can purchase plots of land using cryptocurrency, interact in digital buildings, and engage in various economic and social activities.

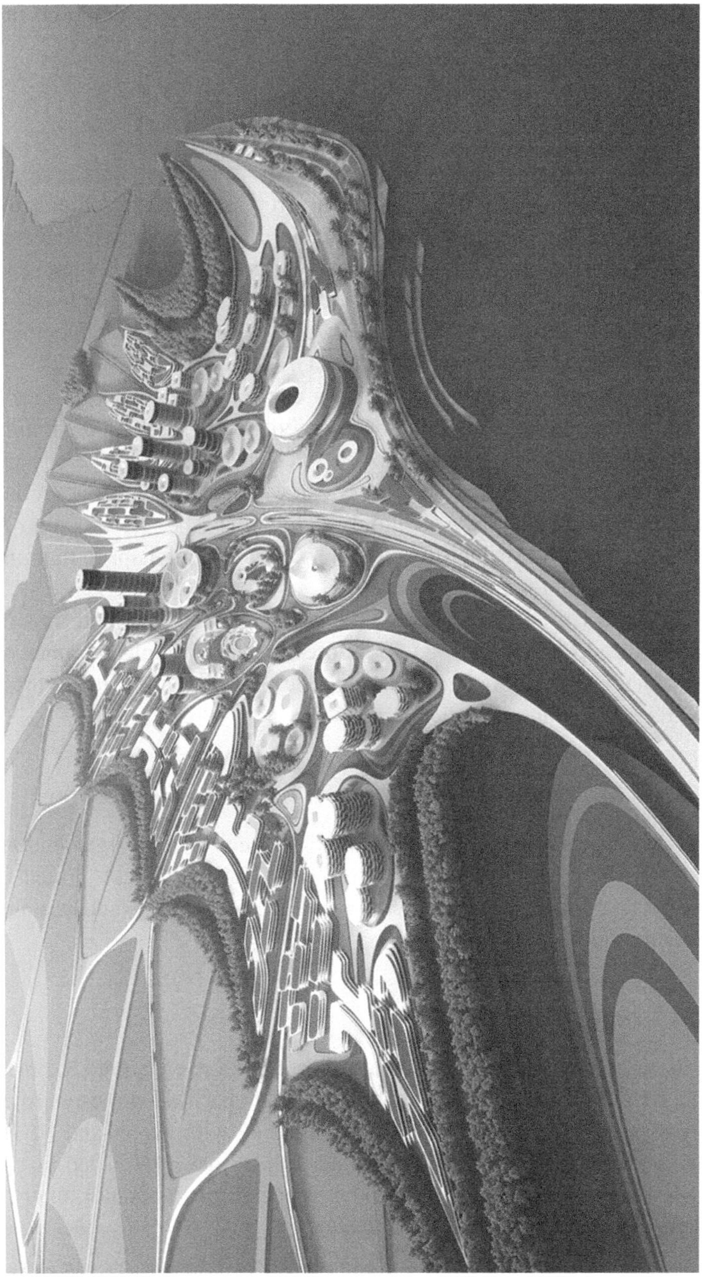

FIGURE 6.1 Overview of Zaha Hadid's design of Liberland.

Inside Liberland's virtual spaces, users experience environments tailored for both work and leisure, seamlessly replicating the operations of a physical city (see Figure 6.2). These environments foster a sense of community and interactivity, providing a testbed for understanding human behaviors and social interactions in virtual contexts (Chow, 2022).

Case 2. Creator's City: The Land of Hope in the Metaverse

Another prominent example of a smart city in the metaverse is *Creator's City*, developed by the Chinese tech giant Baidu. This project represents China's forward-thinking approach to integrating metaverse technologies into urban development. Known as *XiRang*, meaning "Land of Hope" in Chinese, Creator's City is a highly ambitious endeavor (see Figure 6.3). The project is expected to be completed by 2027, thus already in the next 12 months, and already demonstrates remarkable potential in terms of scale and user engagement (Chen, 2021).

The Creator's City sets itself apart with its capacity to host large-scale virtual events. By 2021, the platform could accommodate over 100,000 simultaneous users, showcasing its technological sophistication (Cheng, 2021). The city provides a sprawling virtual environment where users can explore, interact, and collaborate, bridging the gap between the physical and digital realms. The vision behind XiRang emphasizes inclusivity and scalability, creating a shared digital space that promotes innovation, creativity, and social connection.

Baidu's Creator's City underscores the potential of metaverse platforms to revolutionize urban development. Beyond its impressive technological framework, the city aims to offer users a holistic virtual experience, simulating real-world urban dynamics while integrating futuristic elements.

Case 3. Smaller Initiatives in the Metaverse: Fashion and Industry

Among large-scale projects like Liberland and XiRang, several organizations have embarked on smaller, more focused metaverse initiatives. These projects, often spearheaded by major corporations, demonstrate the versatility and scalability of metaverse technologies.

Gucci Vault Land

The Italian luxury fashion house *Gucci* has made a notable entry into the metaverse with its *Gucci Vault Land*. This experimental virtual space allows users to embark on a journey through Gucci's storied history, engaging with games and NFTs (non-fungible tokens) that reflect the brand's heritage. Gucci has established its presence on *The Sandbox*, a metaverse platform, making it one of the first major luxury brands to build a dedicated digital world (Marr, 2022).

FIGURE 6.2 Zaha Hadid's Liberland interior design spaces.

FIGURE 6.3 Overview of Baidu's XiRang metaverse.

Ralph Lauren's Winter Holiday World

Similarly, the American fashion giant *Ralph Lauren* launched a seasonal digital experience in December 2021 on the *Roblox* platform. Known as the *Winter Holiday World*, this initiative allowed users to explore a virtual environment designed to reflect the holiday spirit. Users could shop, interact, and enjoy immersive experiences, further cementing the metaverse as a viable platform for commercial and cultural activities (McDowell, 2021). Putting fashion in the metaverse even inspired the American fashion brand to launch in 2022, one year later, a new corporate sustainability plan called "Timeless by Design", to rethink its impact on the environment and society for the next 50 years.

Siemens and the Industrial Metaverse

In a different vein, Siemens has taken a pioneering approach by introducing the concept of the *industrial metaverse*. Unlike its consumer-focused counterparts, the industrial metaverse aims to mirror and simulate real-world industrial systems, including machinery, factories, buildings, and transportation grids. Siemens envisions this digital counterpart as a tool for optimizing processes, enhancing sustainability, and driving innovation (Siemens, 2024).

Key technologies underpinning the industrial metaverse include cloud and edge computing, industrial AI, and digital twins. By creating a seamless integration between these technologies, Siemens aims to bring practical and tangible benefits to the market more rapidly than traditional metaverse initiatives. The industrial metaverse thus represents a more immediate and functional application of metaverse principles, aligning with industries that prioritize efficiency and sustainability.

The large- and small-scale projects presented illustrate the immense potential of the metaverse as a platform for reimagining urban spaces, commercial ventures, and industrial applications. While large-scale initiatives like Liberland and Creator's City offer glimpses into the future of digital urbanism, smaller projects by Gucci, Ralph Lauren, and Siemens showcase the diverse ways in which the metaverse can transform industries and experiences. Together, these examples highlight the transformative power of the metaverse and its ability to shape the way we live, work, and interact in a digitally connected world.

EXPLORING THE METAVERSE: ITS IMPACT ON HUMAN BEHAVIOR AND THE BUILT ENVIRONMENT

The metaverse, despite its widespread attention and the significant hype surrounding its potential, remains in the nascent stages of development. The primary objective of this emerging technology is to gain acceptance among users while demonstrating its utility for individuals and organizations alike (Markopoulos et al., 2021). Encouragingly, the first indications suggest a positive trajectory. A pivotal moment in the rise of metaverse awareness occurred in October 2021, when Facebook rebranded itself as Meta. This rebranding represents the most significant public relations boost any modern-day technology has received. With Meta's social media platform boasting a user base of 3.05 billion that grows by 5% annually, and 68% of these users logging in daily (Dean, 2024), Meta's promotion of the metaverse profoundly influenced markets and catalyzed trends worldwide.

The Growth of Metaverse Applications

The ongoing trend toward adopting virtual worlds has driven an expansion of metaverse-related projects, investments, and academic research across diverse sectors. The construction industry, alongside the broader built environment, is an early adopter of this transformative technology. This adoption may seem paradoxical, as construction is fundamentally rooted in creating physical structures. However, the industry's embrace of the metaverse to create digital structures before duplicating them in the real world underscores the complementary relationship between digital and physical spaces.

Notable large-scale metaverse initiatives, such as Liberland by the renowned British architecture studio *Zaha Hadid Architects* and XiRang, developed by the Chinese tech company *Baidu*, exemplify this trend. Additionally, smaller-scale projects like *Gucci Vault Land* and Ralph Lauren's *Winter Holiday World* highlight the metaverse's ability to collect valuable insights into user behavior, preferences, and interactions. These insights inform strategic decision-making not only for virtual environments but also for real-world physical spaces.

Metaverse as a Testing Ground for Smart Technology

The metaverse is rapidly evolving from a concept rooted in theoretical exploration to a practical tool for industrial application. This transition enables metaverse platforms to serve as testbeds for smart technologies, offering valuable insights into future adaptations for the physical world (Wang and Medvegy, 2022). For professionals such as architects, civil engineers, building asset managers, and construction project managers, the metaverse provides an environment where client requirements and expectations can be explored with low risk and understood in depth. These requirements are often classified as either functional or nonfunctional, reflecting the user's goals and behaviors (Araiqat and Al-Sartawi, 2023).

The ability to prototype digital representations of physical structures allows professionals to experiment with design elements, spatial arrangements, and functional capabilities. By observing clients interacting with virtual representations of future projects, stakeholders gain a deeper understanding of user needs. This behavioral insight is invaluable for ensuring the final physical structure aligns with the client's vision and expectations.

The Transformation of Physical Environments

The integration of metaverse technologies into architectural practices allows the physical environment to acquire a new level of semiological depth and coherence. In essence, physical spaces are no longer static; they become dynamic entities enriched with virtual overlays and digital interactions. This fusion transforms the built environment into a hybrid space where physical reality is enveloped and infused with digital elements (Schumacher, 2022).

Architectural representation, traditionally confined to three dimensions, has now expanded to include four-dimensional (4D) spacetime elements, thanks to the influence of digital game environments. These 4D forms incorporate dynamic entities such as motion and distortion, which allow for a more comprehensive understanding of spatial relationships and architectural expressions (Uyan & Yazıcı, 2023). By integrating these concepts, architects can create designs that account for temporal changes and user interactions in real time, enhancing the experiential quality of physical spaces.

Gamification and Building Information Modeling

The increasing use of gamified and interactive technologies in the built environment is accelerating the convergence of digital game engines and Building Information Modeling (BIM) software. This integration enables architects and designers to model not only spatial components but also cultural, social, and experiential aspects of a project. As a result, virtual environments are becoming essential tools for evaluating how users will interact with spaces in both digital and physical contexts (Pearson, 2020).

For example, digital game engines allow architects to simulate user behaviors, test design concepts, and refine structural elements within a highly interactive and immersive environment. BIM software complements this process by providing precise, data-driven insights into construction feasibility, material requirements, and sustainability metrics. Together, these technologies form a powerful framework for optimizing both the aesthetic and functional aspects of architectural projects.

Applications Beyond Construction

While the construction industry and built environment professionals are among the early adopters of metaverse technologies, the implications extend far beyond these fields. Virtual environments serve as platforms for social, cultural, and commercial activities, offering opportunities to experiment with new ways of living, working, and interacting. Projects like *Gucci Vault Land* and Ralph Lauren's *Winter Holiday World* illustrate how the metaverse is reshaping industries ranging from fashion to retail. These projects leverage gamification and interactive storytelling to engage users, collect data, and build brand loyalty, demonstrating the versatility of metaverse applications.

The metaverse is reshaping the relationship between human behavior, digital technology, and the built environment. By offering immersive, interactive environments, it provides a powerful tool for understanding user requirements, testing innovative designs, and integrating advanced technologies into architectural practices. From the conceptualization of 4D spacetime representations to the gamified exploration of cultural and social dynamics, the metaverse is driving a paradigm shift in how spaces are imagined, designed, and experienced. As this technology continues to evolve, its influence on the construction industry and beyond will undoubtedly grow, opening new possibilities for bridging the digital and physical worlds.

Sustainable Development Goals in the Context of the Metaverse and Smart Cities

The concept of smart cities in the metaverse transcends the mere application of cutting-edge technologies. It encompasses a broader, more strategic approach that not only focuses on leveraging futuristic interactive technologies but also prioritizes their effectiveness, return on investment, and alignment with global sustainability frameworks such as the United Nations Sustainable Development Goals (UN SDGs). The integration of smart cities into the metaverse offers significant potential to drive positive environmental and social outcomes through their digital operations. By reducing emissions, enhancing accessibility, fostering sustainable communities, driving economic growth, and promoting responsible consumption and production, these virtual cities contribute substantially to the advancement of a range of SDGs (Huovinen, 2022).

One of the primary advantages of metaverse-based smart cities is their ability to create environments that simulate sustainable practices in a digital context, allowing for testing, optimization, and scaling of solutions that can eventually be applied to real-world urban systems. The impact of these digital cities on the UN SDGs can be understood through a layered approach, which categorizes the alignment of smart cities with the 17 global goals, their 169 targets and 232 unique indicators into different degrees of impact. This classification not only demonstrates the direct effects of these virtual cities but also highlights their potential to influence broader societal and economic outcomes.

In the first degree of alignment, smart cities in the metaverse directly and significantly impact almost half of the 17 UN SDGs (see Figure 6.4). These include Goal 9, which focuses on fostering innovation and building resilient infrastructure; Goal 10, which emphasizes the creation of sustainable cities and communities; Goal 8, which aims to promote good jobs and sustained economic growth; Goal 7, which encourages the use of renewable energy; and Goal 13, which calls for climate action. Each of these goals is directly addressed by the functionalities and outcomes of smart cities in the metaverse. For example, digital simulations allow smart cities to test renewable energy solutions, optimize infrastructure, and reduce emissions, all of which contribute to the core goals.

In addition to the first-degree goals (SDG7, SDG9, SDG11, and SDG13), there is also a second layer of indirect impact that smart cities in the metaverse can have on other global objectives. These two goals (SDG5 and SDG10), while not directly influenced by the primary functions of smart cities, are still impacted through the broader societal changes that these cities help to promote. Goal 10, which addresses the reduction of inequalities, and Goal 5, which focuses on achieving gender equality, are examples of how the development of inclusive and accessible digital environments can promote equality in both virtual and physical realms. By providing equal access to services, resources, and opportunities within the metaverse, smart cities can serve as a model for reducing disparities and promoting diversity and inclusion.

Furthermore, the alignment of smart cities in the metaverse with the SDGs does not occur in isolation. These goals can be further combined or expanded upon depending on the specific objectives and strategies of the organizations, industries, and societies

FIGURE 6.4 Smart cities and metaverse alignment with the UN SDGs.

involved in the development of these virtual cities. Organizations may tailor their smart city projects to address additional SDGs that are particularly relevant to their unique context, creating a more customized and impactful approach. For example, a city project in the metaverse that focuses on renewable energy might also incorporate goals related to responsible consumption (Goal 12) or sustainable industry practices (Goal 9). The adaptability of the metaverse as a testing and development space allows for such combinations, enabling greater flexibility in how smart cities can contribute to global sustainability efforts.

Overall, the integration of smart cities within the metaverse offers a promising avenue for advancing the UN SDGs. By aligning digital city projects with these global objectives, organizations and governments can drive significant progress toward sustainability, social equity, and economic growth. The potential of these virtual environments to model, simulate, and optimize real-world solutions further amplifies their capacity to influence the achievement of the SDGs, making them a powerful tool in the global effort to build a more sustainable and equitable future (Huovinen, 2022).

TRANSITIONING FROM THE TRIPLE BOTTOM LINE TO THE QUADRUPLE BOTTOM LINE

The alignment of advanced technological projects with the UN SDGs inherently adheres to the triple bottom line (TBL) theory, also known as the people, planet, and profit (PPP) framework, and goes beyond these traditional considerations (Hubbard, 2009). The direct economic benefits that organizations realize from implementing TBL principles—focusing on profitability, environmental sustainability, and societal well-being—are complemented by indirect advantages stemming from the collection and analysis of data, particularly user behavior and activity within smart cities in the metaverse. This data analysis offers valuable insights to shape the future of the built environment and guide its pathway to be sustainable, efficient, and inclusive.

Moreover, by emphasizing the social and environmental benefits inherent in the TBL model, organizations can not only increase their profitability but also improve their corporate reputation and image. This reputational boost, driven by a commitment to sustainability and positive societal impact, can be seen as an additional, distinct benefit, thus giving rise to what is known as the "fourth bottom line." As a result, the TBL framework, which traditionally encompasses the three Ps—people, planet, and profit—can be extended to include a fourth P—profile. This fourth P serves as a unifying factor that ties together the other three elements, reinforcing the importance of how an organization is perceived in the broader societal context (see Figure 6.5).

This fourth bottom line and its associated advantages are primarily derived from an organization's alignment with the environmental, social, and governance (ESG)

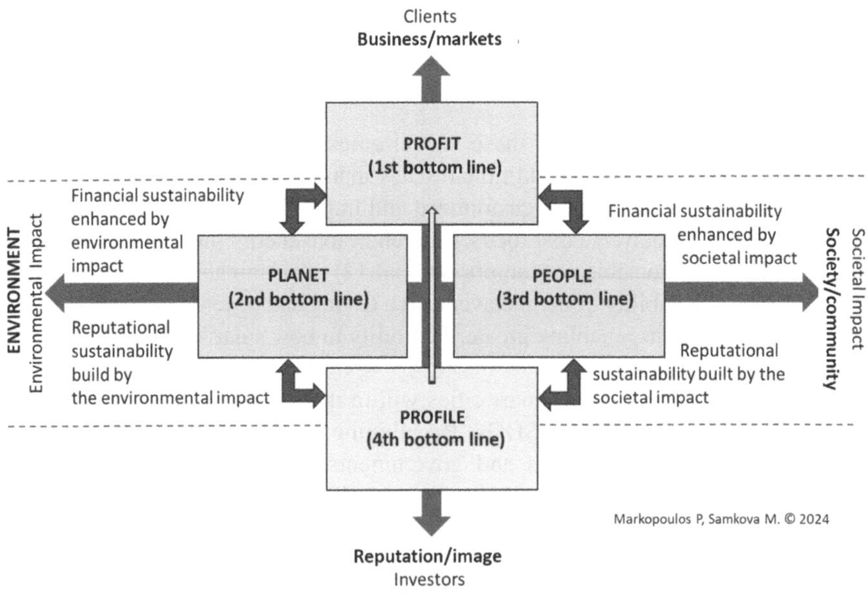

FIGURE 6.5 The quadruple bottom line theory.

criteria and its support for the UN SDGs (Markopoulos & Markopoulos, 2023). These organizations that focus on aligning their operations with sustainability objectives and ESG standards are positioned to not only contribute positively to the planet and society but also to enhance their corporate reputation and increase shareholder and other stakeholders' trust, thus leading to long-term success and growth.

A PROCESS FRAMEWORK FOR SMART CITIES' FUNCTIONAL CHARACTERISTICS AGAINST THE UN SDG AND ESG

The ESG index has increasingly become a critical tool for evaluating corporate performance and managing risks. It plays a significant role in shaping the development and decision-making processes of organizations across various sectors. For companies engaged in the development of smart cities within the metaverse, aligning their projects with ESG criteria is not only beneficial but essential to ensuring sustainability, societal impact, and good governance. These smart city projects in virtual environments can contribute to several ESG pillars, offering both immediate and long-term advantages for companies in terms of both business strategy and reputation.

Table 6.1 provides a concise overview of various smart city projects in the metaverse, linking them to the specific pillars, categories, and themes outlined in the Refinitiv ESG framework (Refinitiv, 2022). This table highlights how virtual smart city initiatives can effectively align with ESG standards, demonstrating that companies can not only meet corporate objectives but also contribute to broader societal and environmental goals through their virtual efforts.

TABLE 6.1
Smart city metaverse projects and ESG requirements

	Pillar	Category	Theme
Smart city projects in the metaverse	Environmental	Emission	Emissions
			Waste
		Innovation	Product innovation
			Green revenues, research and development (R&D), and capital expenditures (CapEx)
		Resource use	Environmental supply chain
	Social	Human rights	Human rights
		Workforce	Diversity and inclusion
			Career development and training
			Working conditions
			Health and safety
	Governance	CSR strategy	Environmental, economic, ethical, and philanthropic responsibility

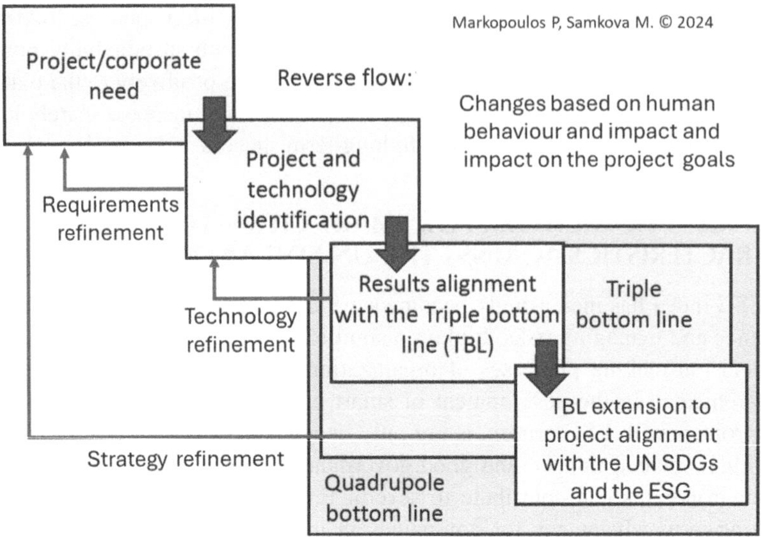

FIGURE 6.6 Value-added process for smart cities in metaverse projects.

By extending the traditional TBL framework to feature a fourth component—the reputational line—organizations strengthen their ESG index scoring potential. This reputational line directly ties to the capacity of smart city projects, particularly those operating within the metaverse, to drive progress toward the UN SDGs and leverage new, cutting-edge technologies. The alignment of these projects with the UN SDGs is essential for ensuring their contribution to creating a sustainable future, while also enhancing the brand images of the organizations and partners involved.

To help organizations navigate this process, an initial framework has been proposed to capture the value-added potential of these projects. This framework is structured in four key stages. The first stage focuses on identifying the specific needs of the project, which often align with the company's broader organizational objectives. The second stage involves defining the project in greater detail and specifying the technologies and tools that will be used to bring the project to life. This stage ensures that the project is both feasible and aligned with the organization's goals. The third stage involves the alignment of the project's outcomes with the established principles of the TBL theory, assessing its impacts on people, planet, and profit. The fourth stage extends this analysis to a quadruple bottom line approach, where the project's results are measured against the UN SDGs and ESG criteria. This ensures that the virtual project adheres to the global sustainability framework while also emphasizing corporate responsibility.

Additionally, a fifth stage in the process provides continuous refinement of the project. This horizontal stage focuses on the iterative improvement of the project at the requirements, technology, and strategy levels, with the goal of maximizing the added value of the project (see Figure 6.6). By refining these aspects throughout the process, organizations can ensure that their smart city projects in the metaverse are not only successful in the short term but also sustainable and impactful in the long run.

LIMITATIONS AND AREAS OF FURTHER RESEARCH

The research conducted for this study has primarily relied on selective secondary research, which provides a theoretical perspective on the potential impact of smart cities within the metaverse on society and the economy, particularly focusing on the human behavior of digital citizens (Luimula et al., 2022). While this approach offers valuable insights into the potential transformations driven by metaverse-based smart cities, it also comes with significant limitations. One of the key challenges faced in this research is the scarcity of completed, real-world smart city projects within virtual environments, which leaves a limited sample for a comprehensive empirical analysis. The lack of extensive, fully operational projects makes it difficult to assess the actual societal and economic outcomes of these digital cities, and as such, comparing their results against the established frameworks of the triple and quadruple bottom lines, as presented in this study, remains a theoretical exercise.

As the field of metaverse-based smart cities continues to evolve, future research will need to focus on empirical data collection from real-life projects and use this data to validate and refine the proposed framework. In particular, a deeper exploration of human behavior within smart cities located in industrial and commercial metaverse environments is necessary. These virtual environments generate valuable data on how digital citizens interact, make decisions, and contribute to the economy and society in these spaces. These data could significantly contribute to the understanding of how the metaverse impacts human behavior in digital cities and, in turn, could allow researchers to test and further develop the proposed framework. Additionally, these findings could provide insights that inform the future development and optimization of smart cities in virtual environments, ensuring that they not only function effectively but also meet the needs and expectations of their digital inhabitants.

CONCLUSION

The evolution of new technologies throughout history has often been accompanied by a surge of optimism, which, in many cases, has not been fully realized. The gap between the expectations surrounding these technologies and the outcomes they actually deliver is frequently tied to the rate at which users adapt to and embrace the technology. The development of smart cities in the metaverse holds considerable promise for transforming urban living, offering innovative solutions for urban planning, sustainability, and social interaction. However, the timeline required for their widespread application is still uncertain, with the key factor remaining the maturity period necessary for users to fully recognize the personal and societal benefits that these technologies can offer.

This research has examined several large-scale smart city projects within the metaverse, including both mega-projects and smaller commercial ventures, all of which share similar technological goals. Despite the differences in scale, these projects are united by their objectives to create sustainable, immersive, and interactive virtual environments that reflect the needs and aspirations of future urban communities. One of the main conclusions of this study is that while the technological infrastructure is being put into place, further emphasis must be placed on incentivizing the widespread adoption and development of these projects. A critical factor is the alignment of these projects with the UN SDGs. By demonstrating the environmental and social benefits of these projects, organizations can secure both public and private support for further development.

Moreover, extending this alignment to the ESG index can provide a pathway for securing the necessary funding and resources required to continue developing and optimizing these metaverse-based smart cities. The integration of ESG considerations into the planning and implementation of these projects not only ensures that they meet global sustainability standards but also enhances their long-term viability by attracting investors and stakeholders who prioritize responsible and sustainable development practices (Markopoulos et al., 2023).

In conclusion, this work highlights the essential relationship between the advancement of technologies like smart cities in the metaverse and the critical need to understand human behavior in these environments. By recognizing and addressing user behavior and engagement, organizations can make these virtual initiatives more sustainable, viable, and impactful in the long run. The future of smart cities in the metaverse depends not only on the technology itself but also on the human beings who interact with and shape these digital spaces. Therefore, further research and development focused on human behavior, user engagement, and real-world data are essential to ensure that these technologies achieve their full potential and are integrated into the physical world in a meaningful way.

REFERENCES

Araiqat, D.M., and Al-Sartawi, A. (2023). The introduction of interior design and architecture into the metaverse and their interrelation to reality-enhancing technologies. In: Hannoon, A., Mahmood, A. (eds) Artificial Intelligence, Internet of Things, and Society 5.0. Studies in Computational Intelligence, vol 1113. Springer, Cham. https://doi.org/10.1007/978-3-031-43300-9_10

Chen (2021). Baidu to expand metaverse offerings and hold virtual conferences via revamped VR app. *Technode*. December 31, 2021. https://technode.com/2021/12/13/baidu-to-expand-metaverse-offerings-and-hold-virtual-conference-via-revamped-vr-app/

Cheng E. (2021). Chinese tech giant Baidu says it could be 6 years before it can fully deliver its metaverse. *CNBC*, December 23, 2021. www.cnbc.com/2021/12/24/baidu-says-it-could-be-6-years-before-it-can-fully-deliver-its-metaverse.html

Chow A. (2022). Zaha Hadid Architects Designs Liberland Metaverse City. Hypebeast. March, 13, 2022. https://hypebeast.com/2022/3/liberland-city-zaha-hadid-architects-metaverse-news

Dean B. (2024). Facebook users and growth statistics. Backlink. February 23, 2024 https://backlinko.com/facebook-users

Finney A. (2022). Zaha Hadid Architects designs virtual Liberland Metaverse city. *Dezeen*. March 11, 2022. www.dezeen.com/2022/03/11/liberland-metaverse-city-zaha-hadid-architects/

Haristianti V, Murdowo D. (2023). The development of the virtual environment and its impact on interior designers and architects. Case study: Zaha Hadid Architects. In Sustainable

Development in Creative Industries: Embracing Digital Culture for Humanities, 296–301. https://doi:10.1201/9781003372486-55

Hubbard, G. (2009), Measuring organizational performance: Beyond the triple bottom line. *Business Strategy and the Environment*, 18, 177–191. https://doi.org/10.1002/bse.564

Huovinen P. (2022). Incorporation of UN sustainable development goals into built environments-related business management concepts. IOP Conference Series: Earth and Environmental Science. 1122, 012057

Luimula M. Markopoulos E., Osterman M., Markopoulos P. Aho J., Ravise W., Saarinen J., Reunanen T. (2021). Avatar Based Multiplayer Functionalities in Next Generation Communication and Learning in Virtual Reality Social Platforms – Case MarISOT Room. Conference: The 12th IEEE International Conference on Cognitive Infocommunications. https://discovery.ucl.ac.uk/id/eprint/10153240/

Luimula, M., Haavisto, T., Pham, D., Markopoulos, P., Aho, J., Markopoulos, E., Saarinen, J. (2022). The use of metaverse in maritime sector – a combination of social communication, hands on experiencing and digital twins. In: Evangelos Markopoulos, Ravindra S. Goonetilleke and Yan Luximon (eds) Creativity, Innovation and Entrepreneurship. AHFE (2022) International Conference. AHFE Open Access, vol 31. pp 115–123. AHFE International, USA. http://doi.org/10.54941

Markopoulos, E.; Luimula, M. (2020). Immersive safe oceans technology: Developing virtual onboard training episodes for maritime safety. *Future Internet*, 12, 80. https://doi.org/10.3390/fi12050080

Markopoulos, P., Markopoulos, E. (2023). Addressing the UN 2030 Sustainable Development Agenda and the ESG Index with Serious Games in Virtual Environments. In: Tareq Ahram and Christianne Falcão (eds) Human Factors in Virtual Environments and Game Design. AHFE (2023) International Conference. AHFE Open Access, vol 96. AHFE International, USA. http://doi.org/10.54941/ahfe1003862

Markopoulos P. Pyae A., Khakurel J. Markopoulos E., Saarnio R., Luimula M. (2021). Understanding How Users Engage in an Immersive Virtual Reality-Based Live Event. Conference: 12th IEEE International Conference on Cognitive Infocommunications (CogInfoCom) (pp. 447–452), 2021. ISBN 978-1-6654-2494-3 https://discovery.ucl.ac.uk/id/eprint/10153242/

Markopoulos, E., Zhao, K., Samkova, M., Vanharanta, H. (2023). ESG and UN SDGs Driven Strategy Generation Process for Green and Pink Oceans. In: Vesa Salminen (eds) Human Factors, Business Management and Society. AHFE (2023) International Conference. AHFE Open Access, vol 97. AHFE International, USA. http://doi.org/10.54941/ahfe1003878

Marr B. (2022). Gucci Enters The Metaverse/ Forbes. Nov. 30, 2022. www.forbes.com/sites/bernardmarr/2022/11/30/gucci-enters-the-metaverse/

McDowell M. (2021) Vogue Business, Feb. 3, Holidays in the Metaverse: What's trending this year. December 8. 2021. www.voguebusiness.com/technology/holidays-in-the-metaverse-whats-trending-this-year

Pearson, L. C. (2020). A machine for playing in: Exploring the videogame as a medium for architectural design. *Design Studies*, 66, 114–143. https://doi.org/10.1016/j.destud.2019.11.005

Refinitiv (2022). Environmental, social and governance scores from Refinitiv. www.refinitiv.com/content/dam/marketing/en_us/documents/methodology/refinitiv-esg-scores-methodology.pdf?elqCampaignId=14314

Siemens (2024). The industrial metaverse. www.siemens.com/global/en/company/digital-transformation/industrial-metaverse.html

Schumacher, P. (2022). The metaverse as an opportunity for architecture and society: design drivers, core competencies. *ARIN*, 1, 11. https://doi.org/10.1007/s44223-022-00010-z

Uyan, B. & Yazıcı, S. (2023). A method for decoding and representing time in four dimensional spaces via digital game environment. *JCoDe: Journal of Computational Design*, 4(1), 99–120. https://doi.org/10.53710/jcode.1248185

Wadham M. (2015). Meet Liberland, the world's newest country. UK Investor Magazine. July 27, 2015. https://ukinvestormagazine.co.uk/meet-liberland-the-worlds-newest-country/

Wang, J and Medvegy, G (2022), Exploration of the future of the metaverse and smart cities. *ICEB 2022 Proceedings (Bangkok, Thailand)*. 12. https://aisel.aisnet.org/iceb2022/12

7 Human-Centered Intelligent Process Control

Vision, Design, and Implementation

Antero Karvonen, Jami Aho, Hanna Koskinen, Mika Luimula, Mari Myllylä, Bastian Tammentie, Susanna Aromaa, Antti Tammela, and Pertti Saariluoma

INTRODUCTION

Throughout the history of industrial operations and process control, new technological tools (e.g., higher levels of automation) have been consistently introduced to enhance efficiency and alleviate the workload and tasks of human operators. Currently, we are witnessing yet another large-scale transformation in industrial operations, driven by a number of simultaneous advancements and societal developments. On the one hand, new societal expectations with respect to the sustainability of industrial processes are increasing. On the other hand, intelligent technologies (i.e., artificial intelligence (AI)-based applications and solutions) and extended realities (XR) that include augmented, virtual, and mixed realities (AR, VR, MR) are transforming ways of working and carrying out everyday process control activities. At the same time, so-called human-centricity, which places worker productivity and well-being as a key issue (Kaasinen et al., 2022), is raising in prominence, particularly in the EU through Industry 5.0 and on a more societal level in Japan through Society 5.0 (Huang et al., 2022). Considering all these aspects at once poses a significant challenge for research, development, and design within the process control industry. They also present challenges for industrial organizations, such as operational transformation, training operative personnel, and building new competencies, as well as maintaining and transferring existing ones.

Industrial process control is a particularly important aspect of the ongoing transformation, because it is the nexus of human–technology interaction (HTI) through which industrial processes are controlled and managed in a concrete sense. This is

DOI: 10.1201/9781003625360-7

where human-centricity and sustainability goals are effectively realized, and the value of intelligent technologies and the industrial metaverse is captured.

One important issue that arises is what kind of questions and goals should be considered in research and development in this emerging context? How can they be integrated with abstract visions and concrete applications? How should design processes and competencies change to accommodate the changing paradigm?

In a previous article (Aho et al., 2024), a research agenda and the associated framework and method that conceptually integrated these different aspects into a holistic approach were described. Here, this framework will be further refined both conceptually and through an examination of case studies within the process control contexts conducted under the outlined research agenda.

BACKGROUND

Technologies such as Internet of things (IoT), AI, and XR have pushed the industry into transitions to a new era that we are now entering (Noble et al., 2022; Xu et al., 2021). One key characteristic of the current Industry 5.0 is its human-centric objective, according to which people should work in a balanced and synergistic cooperation, not in competition, with Industry 4.0 technologies (Xu et al., 2021). Ideally, humans should be able to collaborate with technological agents as "human-aware AI partners" (Korteling et al., 2021, p. 1), supporting each other's activities and strengths in the best possible way (Noble et al., 2022; Rožanec et al., 2022; Xu et al., 2021).

Furthermore, Industry 5.0 is committed to the Sustainable Development Goals (SDG) (United Nations, 2015). Thus, Industry 5.0 should be seen as "a value-driven initiative" (Xu et al., 2021, p. 533) where the core values of industry are shifted from maximizing the scope and progress of technology to increase the well-being and resilience of human individuals, societies, and the planet (Huang et al., 2022; Noble et al., 2022). Similar ideas of a sustainable and a human-centric society are engraved in the concept of a superintelligent Society 5.0, which originated in Japan (Fukuyama, 2018).

SUSTAINABILITY

Sustainability of industrial processes and intelligent technologies is a very problematic topic. As Ripple et al. (2024, p. 1) state, "global fossil fuel combustion and industrial processes account for approximately 90%" of carbon dioxide emissions and other greenhouse gases, which are the primary drivers of the current anthropogenic climate change. Intelligent technologies can have a negative impact on ecological and social sustainability as well due to, for instance, the sourcing and processing of raw materials for equipment and energy consumption associated with data storing, transmitting, and processing or AI algorithms that work on insufficient data or are biased (Vinuesa et al., 2020). However, AI algorithms can also help to make *less*-biased decisions and predict risks more accurately (Kahneman et al., 2021). Intelligent technologies can, for example, improve energy efficiency in manufacturing, production, transportation, and cities, and AI algorithms can be used even to forecast social and environmental phenomena (Bolón-Canedo et al., 2024; Vinuesa et al., 2020).

Well-performing and feasible technologies may also support economic sustainability in terms of providing more value with less cost and use of resources, for instance, by optimizing processes (Vacchi et al., 2021). Consequently, the use of intelligent technologies requires that impacts and risks are appropriately considered in all industrial operations and when designing technology and processes for those operations (regardless of whether it concerns a bigger share of the industrial system or a specific/individual process control tasks). Designers must weigh up what kind of solutions enable the most sustainable end results so that the benefits of technology outweigh any potential disadvantages. Technological solutions themselves must be designed to be as sustainable as possible, from being well-fitting for the diversity of users (Pinzone et al., 2020) and developing energy-efficient source codes, data transfer, and storage systems (i.e., "green AI," Bolón-Canedo et al., 2024), to promoting smart circular economies (Bressanelli et al., 2022).

Opportunities and Challenges

Intelligent technologies and the industrial metaverse hold many opportunities for future industrial work. In the context of industrial operations and manufacturing processes, process control work, and their design, intelligent technologies can be used, for example, to automate repetitive tasks and thus ease the work burden of operators, in quality control, and to generate data syntheses and simulations to improve effectiveness, worker well-being, and safety (Rožanec et al., 2022), while making sure that planetary boundaries are respected (Huang et al., 2022). However, intelligent technology also presents challenges for process control. For example, previous research has illustrated the so-called double black box effect (Wahlström et al., 2024) resulting from the introduction of uninterpretable AI systems to control complex industrial processes. In this instance, not only is the AI system a "black box," but it obscures the target process being controlled, which is also highly complex. "Fragmented intelligence" may occur when data is comprehensible only to experts due to its high complexity or large volume, making the system less usable for non-experts. This phenomenon can be distinguished from "hybrid intelligence" (HI) which has been the vision and goal of our research (see the vision discussion in the next section). Consequently, within the operation of HI mode, at least one critical question arises: How can we ensure that human operators remain in the loop and can maintain an appropriate situation awareness?

In essence, the industrial metaverse is the fusion of digital and physical, where participants can control, monitor, and interact with the industrial processes and systems (EY & Nokia, 2023). Immersive digital spaces offer a combination of social communication, hands-on experiencing, and digital twins (Luimula et al., 2022). Employing metaverse solutions may be closer than often thought for companies with advanced Industry 4.0 solutions (EY & Nokia, 2023). Mirroring real-world machines, factories, buildings, cities, and other large systems to digital spaces creates new possibilities for testing and resource saving. This will further help to innovate sustainable solutions through reduced costs and risks (MIT & Siemens, 2023). However, as the complexity of industrial systems increases, so do skill requirements. Lack of expertise and technical difficulties are pointed out as top challenges in deploying metaverse use cases. Companies also see that the industrial metaverse comes with risks such as cyber and information security and data privacy as the amount of data increases (EY & Nokia, 2023). Moving from existing Industry 4.0 systems thus requires a clear focus on data sovereignty that new blockchain IoT technologies and decentralized connectivity could offer, but this too requires a holistic approach as technical skill requirements increase (Fernández-Caramés & Fraga-Lamas, 2018).

A Framework for Interdisciplinary Coordination

In both the industrial metaverse and intelligent technology domains, a crucial issue seems to be the integration of humans with technologies in a smooth and resilient manner (Kaasinen et al., 2022). Operations should not only undergo transformation but also be grounded in empirically validated contextual facts, such as desirability and acceptability from an operator's viewpoint, and their integration into existing process control workflows. Additionally, they should have a solid scientific foundation, for instance, concerning human cognition. Thus, the problems are multidisciplinary and require interdisciplinary cooperation or collaboration.

The background of the research team responsible for the agenda development has included technology expertise (AI and the industrial metaverse), cognitive science, HTI, human factors, user-centered design, and task analysis. While not exhaustive, this list highlights key disciplines and illustrates both the issues and the group's perspective, which are reflected in the agenda's contents (see next section).

In summary, the complex problems associated with sustainability goals in industry and process control and rapidly developing novel technologies require interdisciplinary cooperation. However, this raises new challenges in creating practical industry applications, fostering a shared language between research fields and industry, integrating research perspectives, and implementing them effectively.

From a Finnish perspective, which provides context for the COACH (intelligent human technology co-agency in process control) project under which this work has developed, the assumption has been that industrial metaverse combined with intelligent technologies can generate a competitive advantage for the Finnish industry when moving from Industry 4.0 toward Industry 5.0. However, a successful transformation requires research and development activities, the management of which can be greatly facilitated by a holistic research agenda, which connects practical realities with research questions and contextual factors.

To this end, we recognized a need for a framework to generate interdisciplinary research agendas that integrate drivers, contextual matters, and normative ideals with research and application in industrial process control (Aho et al., 2024).

THE BIRD'S EYE VIEW: THE FRAMEWORK AND THE AGENDA

The original problem in Aho et al. (2024) was to construct a high-level roadmap for intelligent and sustainability-oriented process control of the future. Research and development roadmaps are useful in providing structure and direction for sometimes long and uncertain R&D processes. Such roadmaps are often highly structured, describing research progress in terms of an envisioned timeline with the use of milestones and parallel thematic tracks. However, when operating at a high level of abstraction with concepts and concrete realities that introduce complexity, it is exceedingly difficult to build realistic and useful roadmaps.

These difficulties caused the problem and solution space to shift toward a more iterative approach, where the function of the roadmap, now understood as an agenda[1], was to facilitate joint discussions where different aspects of the agenda are refined in a resonant manner. This approach has commonalities with the design philosophy described in Schön (2017), where the creative process was described in terms of "fixing" ideas and then refining them in an iterative fashion. The idea was that traditional roadmaps are, in fact, useful, but they can be employed better at a lower level of abstraction, such as when attempting to make progress (and allocate resources) on a single research problem. However, the research problems gain relevance and context at the level of the agenda we described, which means they are not "pure" research problems, but "problems worth solving" from a societal or industrial perspective. Therefore, the structure of a research agenda needs to account for normative visions, current societal and technological drivers, enablers, trends, and application targets – in addition to pure research targets. To realize Industry 5.0 (and related agendas) in practice requires balancing and connecting these different perspectives. This can be described as the overall function of the framework and the research gap toward which it is addressed.

The framework and the process help ensure that all relevant items related to the vision of developing a new intelligent and sustainable process industry are being considered and appropriately addressed. The framework assists in identifying the pertinent research focus, enhancing the validity of research, uncovering underlying issues, and taking more informed steps toward the overarching vision. It also highlights the practical implications for industry, links them to future goals, and examines the trends driving these advancements.

The framework serves a dual purpose. First, it provides an organized method to discuss and create content for research and application paths in emerging technologies. Second, once filled, it enables the formation of reasoned claims and arguments linking research evidence to industrial applications. Additionally, as research and applications advance according to the roadmap, they produce evidence-based knowledge resembling technical norms. As articulated by Niiniluoto (1993), *to achieve X, given that you are in situation B, you should do A*. The framework seeks to emphasize the normative and contextual sides of the formulation to guide researchers toward relevant research problems that can generate knowledge for making valid claims on what should be done at the application level. Thus, the directionality is somewhat different from standard applied science models, where pure research is applied in context.

The essential aspects of the framework are the main categories and their relations; the process of populating it with more detailed contents; and the specific contents as they emerge through the process (i.e., visions, research questions, and application targets impacts). Thus, there is an abstract methodological level, which is open to different emerging technologies and contexts, which will, influenced by the participants and contextual factors, direct the contents in a particular direction. Then, there is the level where the framework has been populated with contents, which allows for a more concrete and iterative discussion on taking the steps toward practice and implementation.

Procedure and Method

The context for the development was in the need to describe in a coherent and relatively simple way the essential outlines of the ideas developed in the COACH project, funded by Business Finland. The purpose was to develop comparable schemas for different projects within a larger ecosystem of research projects connected through the Finnish forest industry's efforts at digitalization.

The development of the agenda began with a temporally structured swimming lane diagram encompassing the categories listed in the next section. However, it was discovered that categorizing research items within a timeline became challenging at the abstraction level of the categories due to the inherent uncertainties in both research and emerging technologies. Therefore, we used a new method by merging temporal–causal elements of a roadmap with a holistic perspective to create a schematic that narratively outlines the research agenda. The idea transformed from a static plan to a structured living document. Philosophically, this approach has commonalities with Quine's (1980, pp. 20–47) pragmatic holism but is extended into a methodological framework and taken outside of the original context of scientific–philosophical epistemology.

Thus, initially, the categories (the blank framework) are best understood as questions or viewpoints, which should be considered, which generate content for a research agenda. Consequently, the agenda develops in a phased manner, from questions to preliminary ideas, and through iteration to a more well-developed agenda. As the research questions and application targets are examined, the agenda moves to a more detailed level – which can influence the items in the agenda but also begins to generate knowledge in the design scientific mode (Niiniluoto 1993).

Categories

The framework consisted of six categories: (1) drivers, enablers, and trends; (2) vision; (3) context; (4) research areas and questions; (5) application targets; and (6) ethics and practical examples of impacts.

The *first* category makes visible emerging technologies, societal trends, and other relevant factors identified in the environment. These can be understood as influencing factors or factors that enable (or force) novelty in the context of research and application and the concrete context. The *second* entails taking a normative stance with respect to a desirable future state. This statement summarizes and scopes what the relevant research targets and application areas are. The *third* makes visible relevant aspects within the context of research and application. What the context entails cannot be exhaustively defined, but even so it allows for raising issues regarding both particular and general features of the domain of research and application. The *fourth* lays out the basic directions where research should be geared toward, given the overall context of the total roadmap. The *fifth* connects research areas and questions with practical industrial problems and applications. While the entire roadmap is geared toward normative and contextual applied science, this part should articulate

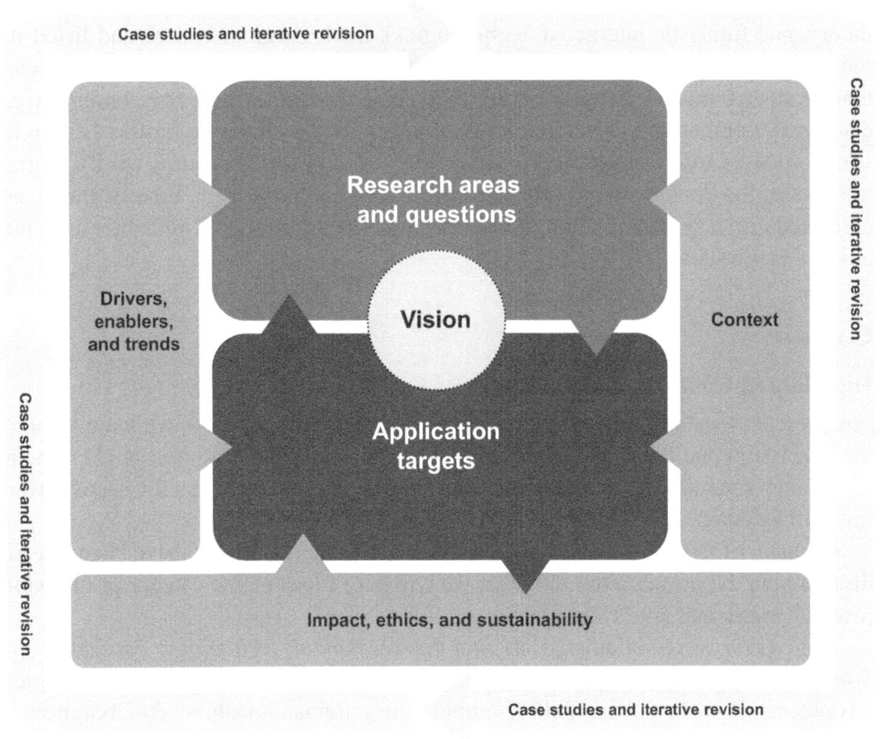

FIGURE 7.1 Categories and their relationships are visualized in the framework, illustrating an iterative revision process via case studies. (Figure by the authors.)

the ways in which, for example, industrial work or design practice can benefit from research. The fourth and fifth categories require a strong iterative approach between them. The *sixth* category makes the case that the overall research agenda leads to beneficial outcomes when applied in practice and includes ethical considerations. Figure 7.1 visualizes the categories and some of their relations.

Participants

The participants comprised a multidisciplinary group of researchers, featuring experts from cognitive science, human factors, and fields focused on the design of advanced interactive technology. Seven core participants were involved in the agenda generation process, and occasionally, other researchers from the project also contributed to the work. Although the contents of the agenda were informed by data obtained from empirical case studies, such as interviews with operators and designers, the interviewees themselves did not take part in creating the research agenda.

The agenda development began in early 2023. Initially, the construction began by dividing the researchers into three groups, each group approaching the six categories from their own research approaches (i.e., cognitive science, practice theory, and futuristic interactive technologies). Each group discussed and listed the most important factors or items related to each given category. After this, the content produced was reviewed together in several sessions, where the contents were discussed, compared, evaluated, and finally modified to be more uniform and relevant for the COACH research project's context (i.e., process control work). During this phase, the contents were evaluated against research findings. Finally, they were compiled into a synthesis summarizing the main ideas, concepts, and objectives into a list of key issues.

Contents

Highlighted Contents and Findings

For present purposes, it is sufficient to raise a few highlights and explicate some of the discussion and thinking behind them. Overall, the content of the agenda remains on a rather general level and thus requires further empirical work; a few steps toward this will be taken in the case studies section of this paper.

As many of the drivers and trends, as well as contextual issues, have been already discussed in the introductory remarks, we will here focus on the categories of vision, research areas, and application targets.

The *vision* was articulated as *intelligent human technology co-agency for future sustainable industry operations*. This vision aims to integrate technological advancements with sustainability, emphasizing human–machine collaboration in industrial process control. Co-agency emphasizes the importance of supporting and sustaining human agency and action, while acknowledging that intelligent technology can influence this by performing autonomous activities and occasionally taking over certain tasks. This perspective highlights the necessity of simultaneously designing operations, work processes, tasks, and technology. This echoes the holistic and pragmatic philosophy guiding the framework as well.

Consideration of both drivers and trends, as well as contextual issues, gave a sharper background to relevant *research areas and questions*. For example, the cooperation and co-design of human-AI systems imply research questions related to theory and design languages (Saariluoma & Karvonen, 2023), as well as the scientific basis and approach for designing interactions (Howes, Jokinen & Oulasvirta, 2023). These are crucial for placing design and implementation on a rational basis. Thus, one thematic area that emerged was the *relevant scientific knowledge and practice* for realizing the agenda. On a *methodological level*, information engineering of industrial systems can be enhanced using cognitive mimetics (Karvonen et al., 2023), which allows operators' mental contents, information processes, and actions when interacting with technologies to be detailed and computationally modeled as human digital twins (HDTs) (Saariluoma, Karvonen & Sorsamäki, 2021; Saariluoma et al., 2024). Hence, HDTs serve as middleware (Kaarlela et al., 2023) that integrate not only physical but also human cognitive attributes into the virtual realm. Furthermore, to facilitate a more systematic approach to *design work*, it's essential to examine emerging issues in design systems (DSs) for intelligent technology and general design methodologies (Myllylä, Karvonen & Koskinen, 2024). *Studying and testing human interactions* with new intelligent technologies like AI assistants and robot systems, especially in digital workspaces like the industrial metaverse (Kaarlela et al., 2023), emerged as another key thematic area. This complements iterative and exploratory design methods involving various tasks like initial task analysis (Norros, Savioja & Koskinen, 2015), actor-network and value analysis (Kaasinen et al., 2022), and planning test setups and developing new interfaces and virtual platforms (Kaarlela et al., 2023).

Expert networks, changing operational models, and the metaverse issue further research and application issues. Future industrial work may involve a much more diffuse collaborative model, particularly in process control and engineering, whereby relevant expertise is brought to bear on problems on an on-demand basis, possibly globally. This change is driven by the general advance of digitalization, which involves the convergence of many technical and conceptual phenomena, which generates complexity. This creates both challenges and opportunities. As AI technologies advance in complexity and become more opaque while becoming more important for process control, there may be a reduction in local knowledge of the process (Wahlström et al., 2024), which will still be important due to contextual and contingent reasons. Understanding the effects particular forms of intelligent technology can have on deskilling or process understanding is crucial. The industrial metaverse and the new ways of working and learning that it enables could result in a more collaborative paradigm in terms of training, design, and operations.

At the end of the day, *successful implementations and their impact* will be the decisive factor. Intelligent technologies will affect industrial processes, from daily tasks on the factory floor to broader organizational structures and management. These impacts may change how work is performed and the skills and competencies that are required, influencing how organizations plan and manage the integration of these technologies to keep work engaging and acceptable.

Digital and virtual intelligent applications enhance operational efficiency and quality by simplifying the management, understanding, and prediction of complex human–cyber–physical systems. Therefore, design thinking and methods for developing intelligent technologies must be continually updated to capture new benefits and mitigate risks. However, while our findings indicate that both intelligent technology implementation and novel remote and decentralized work processes have generated benefits, challenges also remain.

The work on the framework has shown that the current industrial transformation is a complex phenomenon, which requires a multidisciplinary research agenda and collaboration, not only between industry and research, but also between different scientific research viewpoints. Collaborating with industry in research can create more resilient, adaptive, and sustainable organizations and processes through co-designing technology and human operations. This approach can boost investments, productivity, and well-being. A multidisciplinary, multistakeholder agenda is essential, and the outlined framework seeks to apply these ideas to a collaborative research agenda drafting.

However, creating a plan for gaining a grip over rapidly evolving technologies is difficult. The items on the agenda can be examined from various perspectives, ranging from broad, comprehensive views to more granular, detailed analyses. This approach may also raise new questions about understanding interconnections and viewing them as part of an integrated system. Therefore, it is essential to alternate between the overarching view and the finer details when discussing and refining the agenda. Next, a selection of the case studies conducted within the COACH project will be described.

CASE STUDY REFLECTIONS

To take the agenda further and validate its content, it is necessary to reflect it against empirical evidence. In this section, some reflections will be provided, based on the case studies conducted. The case studies are still mostly works in progress and thus will not be presented in much detail. Rather, they are treated in terms of emerging issues and findings, particularly in terms of the research agenda and as a way of demonstrating the way in which a dialog can be established between the two levels.

Study 1: Intelligent Technology in Process Control

This case study was related to the design and implementation of a predictive AI-based service for the wastewater treatment plant of a cardboard factory. The new service is aimed at providing an improved overview of the process as well as giving suggestions about chemical dosing.

During the development and implementation process of the AI-based service, its two designers were interviewed three times: beginning of the development, piloting phase, and after six months of use. The interviews progressed gradually from discussing goal setting and user requirements to getting feedback about the new service in use. In addition to designers, two wastewater treatment plant operators (users) and a production engineer were interviewed, and their work was observed. The first interview was prior to the implementation of the service, and the second time was after the implementation when they had been using the system for a few months. The first interview concentrated on operators' work and process control tasks, whereas the second interview focused on the user experiences of the new service. Additionally, a questionnaire including 24 questions related to user experience and usability was introduced in the second interview round.

Findings

The results show that there is potential in integrating intelligent technologies in process control but also highlight the importance of carefully considering the HTI and the need for better integration of expert users' experiences and knowledge into the AI system. Findings indicate that there are still many challenges that should be addressed in AI services regarding, for example, usability, information availability and understandability, reliability and validity, interaction and collaboration, and trust.

Regarding the proposed framework, this use case was related to the drivers such as AI technologies and Industry and Society 5.0 and to the human-AI interaction research area. In this study, it was noticed that the AI-based service could be designed in such a way that its usability was at a good level, and it worked well in process control work. Despite this, operators felt that they could not always trust the values suggested by the service or that they could not interact with the service (i.e., bidirectional information flow). As a result, the system was unable to receive all the tacit knowledge that operators would have wanted to share with the service. Based on this, for future research on intelligent systems, the inclusion of contextual and tacit knowledge would be important. In other words, the design of intelligent systems should place particular emphasis on the active participation of users throughout the design process, so that tacit and contextual information can be integrated into an intelligent system. Thus, this case study validates some of the contents in the research agenda, but also introduces new more specific issues to be researched further.

Several findings related to design also emerged through this study (see Myllylä, Karvonen & Koskinen, 2024). The research showed that despite the novelty of the technology, the issues are not necessarily altogether new, but rather repeat already well-established findings on, for example, basic usability and HCI design. Thus, there is a clear need for systematization of knowledge, not only new research findings, but also established knowledge so that intelligent technology application does not stumble upon basic mistakes, which could be avoided. One approach to tackle these issues is through DSs, which ideally serve as a coherent and systematic repository of design-relevant knowledge and assets, through which the development of design can progress (Myllylä, Karvonen & Koskinen, 2024).

However, intelligent technologies also introduce new issues for which there are no well-established design patterns. For example, only preliminary ideas are available for solving the double black box phenomenon outlined in Wahlström (et al., 2024). Consequently, this requires that design methods specifically related to intelligent technologies are used and developed further. Methods such as core task analysis (Norros, Savioja & Koskinen, 2015), content-based cognitive analysis (Myllylä and Saariluoma, 2022; Saariluoma, Cañas Delgado & Myllylä, 2024), and others (Oulasvirta, Jokinen & Howes, 2022; Saariluoma, Cañas, and Leikas, 2016) can help to collect information, which will be relevant for modeling human activities in specific contexts. Overall, the conclusions drawn from case studies on these design-relevant topics highlight their importance within the general research framework.

Study 2: AI Assistants and Metaverse Applications

These case studies involve the introduction of a novel service concept in the case company, which is intended to provide various kinds of advice to the customers related to the processes that they operate. As the company provides many kinds of services (e.g., technical systems, process advice), the novel service concept requires pooling the company's expertise and finding ways of applying it on an on-demand basis. The studies both involve the application of AI, the second in combination with a metaverse approach.

Study 2 A: AI Assistants

The operators' work is currently hindered by the underutilization of new systems, inconsistent technical standards, and communication challenges with customers. The goal of this case was to study the use of AI in the work of remote operators when they assist their customers.

To conceptualize the research problem relating to human-AI interaction and co-agency, a five-part schema was developed (Figure 7.2). This schema was developed in a research workshop where we considered how we could methodologically study how AI can support the human worker.

The schema in Figure 7.2 demonstrates five different stages (or modes) of problem-solving with AI. In the *human-only (HO)* stage, the problem is solved by an individual human or two or more humans working together. In the *HI* stage, the problem is solved by humans with assistance from an AI system. In the *AI-only (AO)* stage, the problem is solved by an unassisted AI with a human worker reviewing and approving the solution. This case study focuses on the HI stage in attempting to find ways for AI-assisted problem-solving in the case company environment.

Training the AI personas was done by interviewing six (6) Finnish and seven (7) international remote operators involved in the new service concept about their work. The topics discussed in the interviews included the used problem-solving procedures, available tools, current challenges, and future development needs. The obtained interview material and material relevant to the operators' area of expertise were used to train the GPT (generative pretrained transformer) AI personas. The GPT's answers were limited to being based on the provided anonymized interview and background material only, significantly reducing the likelihood of "hallucinations" (i.e., counterfactual content (Bryce et al., 2024, p. 235)). The GPT personas were prototyped using ChatGPT, as has also been explored in similar studies (Cheng et al., 2023; Fernandes et al., 2024).

Data freely obtained from the company via the internet were also used to train AI. Such information was, for example, the working model of remote operators announced by the company. Information about the company's staff, for example, number and education, was also used. These interview-based GPT personas are trained not only with general task or area-specific instructions (e.g., the operations of a board and paper machine) but also with information derived from the skills and knowledge of real human operators. They are "digital shadows" of human operators, capable of adding otherwise unwritten human know-how to the suggestions of the personas.

The GPT personas were tested by analyzing the interviews with the help of the AI (hallucination control) and by asking the personas, for example, about the educational background of remote operators or their way of approaching problem-solving (content comprehension test). The analysis also included the process by which the operators act when solving customers' problems.

A key discovery about the AI assistants was that each interview resulted in a slightly different AI assistant. This was tested by giving them different problem situations to solve. These included, for example, "a sudden increase in alert notifications from multiple customer sites indicating potential equipment failures" or "environmental

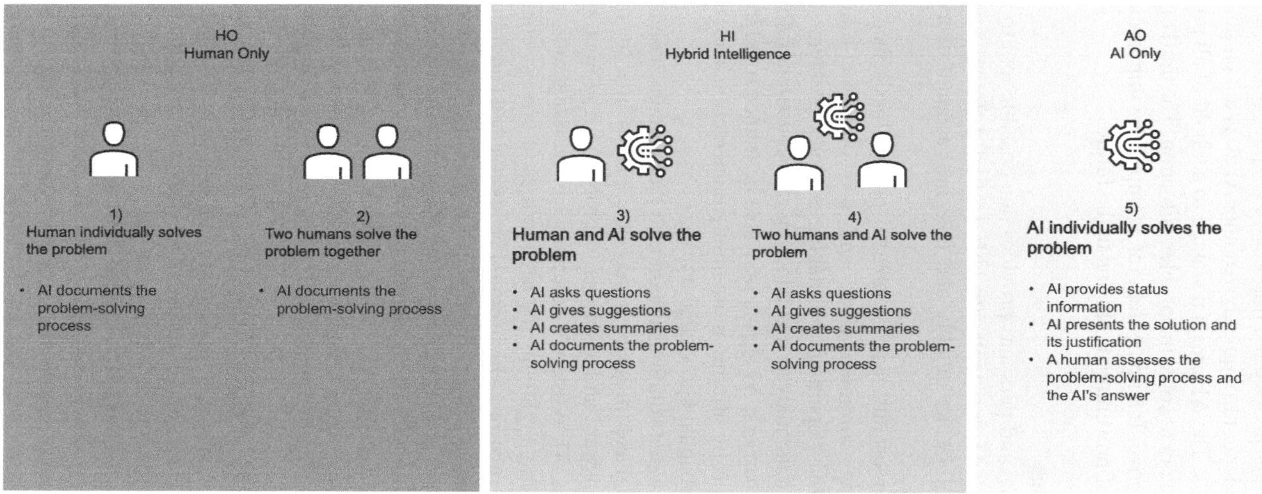

FIGURE 7.2 Five modes of problem-solving with AI. (Figure by the authors.)

compliance requirements change, necessitating immediate adjustments in emissions controls at customer sites."

Each AI assistant went about solving these situations in slightly different ways. One of the AI assistants stressed the importance of quick action. It wanted to get the problem situation under control quickly. Another AI assistant emphasized the importance of measurements. The AI assistant wanted to get hold of all possible data at first and only after that set out to solve the problem situation. The third operator could start with a bigger perspective. It emphasized, for example, the importance of information and the importance of preventing problems in the future.

Preliminary Findings

Training AI assistants and personas provides one possible tool to support the work of remote operators. AI assistants are able to find different options for solving problem situations; they act quickly and take into account everything they have been taught.

The GPT-based assistants may be able to provide assistance by standardizing responses, improving communication, and providing real-time, context-aware support. They can assist operators by making complex problem-solving more efficient through better data analysis and handling of inconsistent information and help overcome language barriers. AI agents may also be trained to address the underutilization of new systems by offering more intuitive interaction and guidance and reducing the reliance on manual knowledge transfer.

Nevertheless, AI-based solutions continue to face certain challenges. AI assistants may hallucinate and make mistakes. It may take a lot of time for operators to check the correctness of answers and solutions to problems (Figliè et al., 2024). This is essential because the AI assistant's solutions can be dangerous and costly due to potential mistakes. This highlights the peculiarities of process control as the site of the industrial transformation, which should be approached carefully in a holistic and research-oriented manner. Mitigating this challenge is not only a technical problem, but also an operational and strategic one, which shows the importance of a holistic approach when dealing with emerging and transformative technologies such as large language model (LLM)-based AI (see, e.g., Bryce et al., 2024).

Additional challenges can be identified, particularly in integrating AI into complex human-AI collaborative environments: The operator's potential loss of tacit knowledge as AI systems become more involved in decision-making processes must also be considered as a potential risk (Fernandes et al., 2024). The human knowledge-based AI agents as presented in this experiment aim to bridge that gap by incorporating the lived experiences and expertise of workers, ensuring that human knowledge continues to play a vital role in hybrid systems.

In the future, AI assistants will also need other data in addition to premade training. Information about how previous similar problems have been solved in the company in question is of primary importance. The AI assistant learns to solve the problem through a successful prior solution. The described hybrid approach offers the advantage of incorporating human understanding through structured interviews, alongside other data and clearly defined problem-solution patterns.

These AI personas also offer possibilities for designing technologies and operations. The same models could be adjusted to provide an interactive and dynamic way for designers to take different kinds of actors into consideration when designing new technologies for industrial use or when designing new operational models. Such development could lead to powerful new additions to DSs (Myllylä, Karvonen & Koskinen, 2024). In terms of the framework and the research agenda, this makes it apparent that with relatively small adjustments, research-oriented case studies can contribute to different application targets (in this case, AI assistants supporting operational problem-solving and designers' work). This illustrates the value of maintaining a holistic research agenda in a "living document" mode that can be used to reflect on more general findings and discover novel connections beyond the details of specific research processes.

Conceptually, AI assistants and methods used in building them generate new dimensions for the development of HDTs (Saariluoma, Karvonen & Sorsamäki, 2021; Saariluoma et al., 2024, p. 2), which were outlined in the research agenda. Novel opportunities for enriching HDTs and connecting them with concrete environments are foreseeable through the industrial metaverse, as the emphasis there on building an interface between digital and concrete reality provides tools for capturing human actions.

Study 2 B: AI-Assisted Metaverse Meeting Room

This case study explores opportunities with metaverse meeting rooms in industrial contexts such as process control and problem-solving that may involve interaction with, for example, operators and process specialists. The goal was to develop and test solutions where AI systems assist meetings by providing information to enhance discussions.

During the development, a collaborative metaverse environment was created with Unity3D and Proverse platform, where users can join as avatars and communicate through voice and chat systems. This development work required collaboration from both graphics and coding, but also feedback from researchers, industry partners, and end users. The environment development included functions to upload 3D models and other contents for collaborative viewing as well as using OpenAI LLM to assist in note creation for these objects. With the recent innovations such as ChatGPT and project focus on human–machine collaboration, an additional need was identified. How can AI assistants be included as a collaborative partner in a problem-solving case within a metaverse context? Thus, this case study followed a variation of HI mode 4 (see Figure 7.3).

Using OpenAI's services, an AI assistant was created. Its main purpose is to listen to and give additional suggestions to the participants during their meetings based on the topics discussed. The system was built so that assistants can be added and edited from OpenAI website, and they will show up within the Proverse client application. When starting a meeting, an assistant is chosen who will follow the instructions and prompt what has been given. As the participants communicate through voice, a text-formatted chat log is generated using a speech-to-text service called Whisper. Recent messages are then sent to the LLM based on settings such as "Suggestion Interval,"

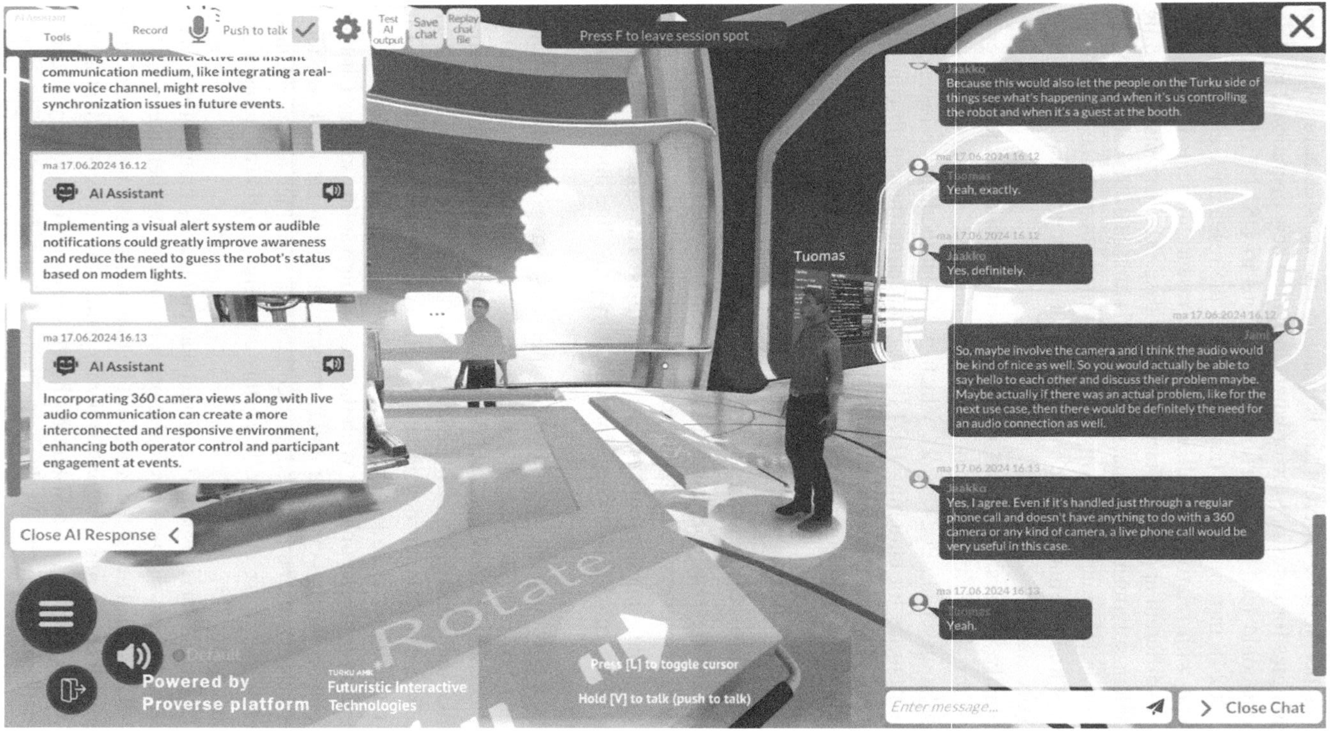

FIGURE 7.3 Automatic AI Assistant giving suggestions to participants during a meeting. (Screenshot taken by authors.)

which decides how many messages should occur before requesting a new suggestion from the AI or "Recording Length," which decides how many recent messages from chat log will be sent to the AI. Meeting participants will then get AI-assistant suggestions based on these parameters.

Preliminary Findings

This case study is still in progress, but early results show that these types of automatic assistants can give relevant, valuable input and contribute to the overall discussion. Additional tools that use prompting for various needs, such as creating the meeting notes, next steps, and other more targeted requests, are effective ways to reduce work. The study produced performance testing of this system for different AI assistant suggestion intervals. This interval for the AI suggestions was then used for the automatic assistant in the group discussions. Then, data were gathered from students on usefulness, satisfaction, and ease of use with the usefulness, satisfaction, and ease of use (USE) questionnaire as well as determining situational load during the group discussion.

Regarding the proposed framework and agenda, this use case was related to the drivers such as AI technologies, industrial metaverse, HDT, Industry 5.0, and Society 5.0 and also to several of the research areas. Some initial observations from this study highlight design issues, as some testers were unclear about the intended use of this AI assistant. Many assumed it would function as a chatbot that directly answers questions and were initially unsure of its purpose before understanding the concept. This can be interpreted in terms of both AI design and a challenge in terms of implementation. While explicit instructions can provide solutions, it is crucial to acknowledge that users come with preconceptions. Therefore, system design and training must carefully consider these expectations. This shows both the importance of integrating well-established HCI knowledge into novel technology design and generating contextual understanding of the users in question (see also Myllylä, Karvonen & Koskinen, 2024, as well as case study 1).

Other notes include participants feeling relief when they could speak in their own language, which shows the need for diverse language handling capabilities in multilingual networks and fits in the human technology and metaverse research areas. When such technologies are developed for global use, where many cultures and languages may interact, these questions become increasingly important. Unfamiliarity with the given topic was something that also affected the discussion depth, as could be seen in how the students sometimes struggled to participate in the discussion. This can be seen as part of human-AI interactions research and is an important aspect to consider when applications are targeting different expertise levels such as managers, operators, and other end users. Expert opinions are still expected during this project to reflect on the data collected from students and will give a good insight into how their knowledge of the topic reflects on the suggestions given by the AI assistant.

Several future research questions have emerged. How can AI assist in solving challenges over a longer period that involve multiple meetings? How can intelligent systems proactively keep track of meeting topics and discussions and point out previous findings from current and previous cases to human participants effectively and

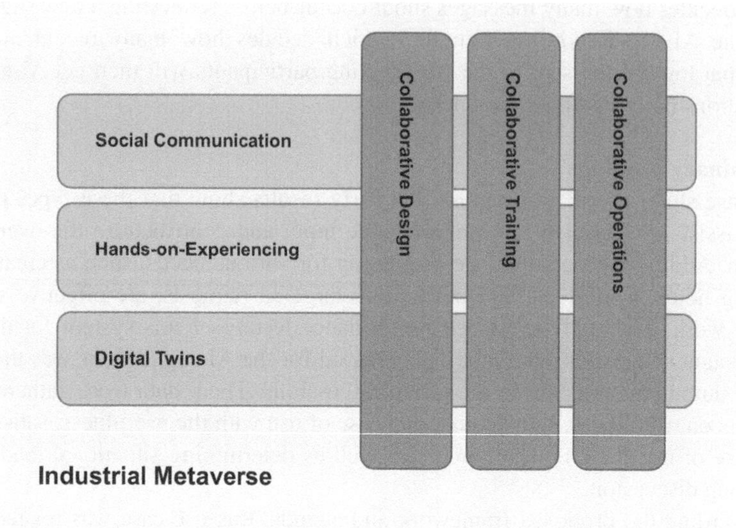

FIGURE 7.4 An extended definition of the industrial metaverse. (Figure by the authors.)

accurately? Additionally, sensor data from factory tags could be introduced to these systems as well to present users with relevant data. While prototypes like those in this study can test certain aspects and functions, integration into existing industrial systems would give insight into actual usage and user experience of such systems. This provides validation for the emphasis in the framework on applications and concrete impact.

On a more general note, based on a previous definition (cf. Luimula et al., 2022), and experiences collected from various RDI (Research, Development and Innovation) projects, the definition of the industrial metaverse should be extended to cover collaborative elements. First, the industrial metaverse enables communication and collaboration on a level not achieved using present platforms (Kontio et al., 2023). Second, there is a clear need for *collaborative training* solutions, such as when a team of individuals needs to learn to respond to acute and hazardous events. In this chapter, the second element of collaboration in our definition (Figure 7.4), namely *collaborative design* is introduced in detail. While the case described focused on collaborative problem-solving, this naturally applies to design operations as well. Additionally, the metaverse allows for the integration of point clouds with 3D models of buildings and structures, providing new visual and interactive methods for design. Thirdly, collaborative operations within industrial metaverse facilitate complex or potentially hazardous industrial tasks by using immersive, photorealistic multiuser environments with situation-aware user interfaces (Nokia 2024). These collaborative solutions require social communication, hands-on experience, and integration with the real world through digital twins as previously described.

CONCLUSION

The case studies reveal detailed insights and specific research questions related to concrete industry applications. Thus, there is a necessary divergence between the two levels of abstraction of the agenda and the case studies, which should not be forcefully integrated or slavishly followed. Rather, the opportunity is in establishing a rich interplay between the "bird's eye view" through the framework and the practical use case studies to enable the discovery of general patterns of development and give opportunities for cross-disciplinary cooperation to make meaningful transformations. This is crucially important, as emerging technologies and their applications in the process control context are laden with novel patterns that are presently conceptually unrecognized and under-researched.

Although our case studies have proceeded in a somewhat discrete manner, they validate the premise of the framework that the key issues related to intelligent technology, metaverse, and human work and action are always interlinked in practical applications. Though obvious in theory, our case studies show this is not always achieved in industrial R&D. Furthermore, the emphasis placed on establishing a strong relationship between research and concrete industry application appears to be sound, which strikes a balance between practice-driven and academic research.

The framework encourages a collaborative approach to industrial transformation in process control context. Not only at the level of drafting a research agenda, but also our case studies reveal as a common thread the need to engage with stakeholders in developing transformative technologies. In principle, both the industrial metaverse and intelligent technology can further advance this idea through collaborative design, learning, and operations. Intelligent technologies, like AI assistants, can be developed and designed to reflect and support people, rather than replace them. Case studies and the conceptual model of HI (see Wahlström et al., 2024) and modes of human-AI interaction suggest that even small details in their implementation can significantly affect usability, trust, and acceptability. These factors influence whether the technology is adopted and if its potential value is realized. This is one of the reasons why academy–industry collaboration in a multidisciplinary mode is valuable.

While collaborative development is not always easy or straightforward, the findings illustrated in this chapter suggest that with proper methodology and frameworks, there are no principal obstacles to achieving it. The vision articulated in the framework – intelligent human technology co-agency in process control – is revealed to implicate a comprehensive collaborative model. Technology should either collaborate with humans intelligently or enable human collaboration. There is a need for research and development models that consider specific circumstances while linking to general patterns in transformative technologies and scientific understanding.

The core focus in the presented work has been on process control and expert support delivery as well as system design. The viewpoints so far articulated only scratch the surface, as, for example, theories of organizational change or innovation management and leadership recognized in the overall agenda present an obvious point of further research (Aho et al., 2024). The crucial idea is that by establishing a link with case examples and maintaining an open and active mode in the agenda development, such extensions are not only possible, but natural.

Thus, the model of research and development for transformative technologies embodied in the framework and illustrated in the case studies can be called contextual and collaborative applied science. For intelligent technology and the industrial metaverse, this appears to be a latent factor that should be emphasized across research and application to realize a successful transformation of industrial process control.

ACKNOWLEDGEMENTS

This work has been funded by Business Finland through the COACH -project (project number 3289/31/2022). We thank the case companies and colleagues in the project for their input.

NOTE

1 The original paper used the term roadmap in an inconsistent manner. Here, this is corrected by using the main concepts *agenda* and *framework*. Agenda refers to the framework populated with contents using the methodology described.

REFERENCES

Aho, J., Karvonen, A., Koskinen, H., Luimula, M., Myllylä, M., and Tammentie, B. (2024). Human centered research agenda for intelligent technology in process control. In: Beata Mrugalska, Waldemar Karwowski, and Stefan Trzcielinski (eds) *Human Aspects of Advanced Manufacturing, Production Management and Process Control*. AHFE (2024) International Conference. AHFE Open Access, 146. AHFE International, USA. http://doi.org/10.54941/ahfe1005159

Bolón-Canedo, V., Morán-Fernández, L., Cancela, B., and Alonso-Betanzos, A. (2024). A review of green artificial intelligence: Towards a more sustainable future. *Neurocomputing*, 599, Article: 128096. https://doi.org/10.1016/j.neucom.2024.128096.

Bressanelli, G., Adrodegari, F., Pigosso, D.C.A., and Parida, V. (2022). Towards the smart circular economy paradigm: A definition, conceptualization, and research agenda. *Sustainability*, 14(9), Article: 4960. https://doi.org/10.3390/su14094960.

Bryce, C., Kalousis, A., Leroux, I., Madinier, H., Pasche, T., and Ruch, P. (2024). Exploring the dual role of LLMs in cybersecurity: Threats and defenses. In: Kucharavy, A., Plancherel, O., Mulder, V., Mermoud, A., and Lenders, V. (eds.), *Large Language Models in Cybersecurity: Threats, Exposure and Mitigation* (pp. 234–242). Springer. https://doi: 10.1007/978-3-031-54827-7.

Cheng, K., Li, Z., Li, C., Xie, R., Guo, Q., He, Y., and Wu, H., 2023. The potential of GPT-4 as an AI-powered virtual assistant for surgeons specialized in joint arthroplasty. *Annals of Biomedical Engineering*, *51*(7), pp.1366–1370.

EY & Nokia. (2023). The metaverse at work: New study in partnership with EY around industrial and enterprise metaverse. www.nokia.com/metaverse/industrial-metaverse/the-metaverse-at-work-research/

Fernandes, D., Garg, S., Nikkel, M., and Guven, G. (2024). A GPT-powered assistant for real-time interaction with building information models. *Buildings*, *14*(8), p. 2499.

Fernández-Caramés, T. M., and Fraga-Lamas, P. (2018). A review on the use of blockchain for the Internet of things. *IEEE Access*, 6, 32979–33001. https://doi.org/10.1109/ACCESS.2018.2842685

Figliè, *R.*, Turchi, T., Baldi, G. and Mazzei, D (2024). Towards an LLM-based Intelligent Assistant for Industry 5.0. CEUR Workshop Proceedings. Proceedings of the 1st International Workshop on Designing and Building Hybrid Human–AI Systems (SYNERGY 2024), Arenzano (Genoa), Italy, June 03, 2024.

Fukuyama, M. (2018). Society 5.0: Aiming for a new human-centered society. *Japan SPOTLIGHT*, July/August 2018, pp. 47–50. www.jef.or.jp/journal/pdf/220th_Special_Article_02.pdf (Accessed 11 October 2024).

Howes, A., Jokinen, J. P. P., and Oulasvirta, A. (2023). Towards machines that understand people. *AI Magazine*, 44, pp. 312–327. doi:10.1002/aaai.12116.

Huang, S., Wang, B., Li, X., Zheng, P., Mourtzis, D., and Wang, L. (2022). Industry 5.0 and Society 5.0—comparison, complementation, and co-evolution. *Journal of Manufacturing Systems*, 64, pp.424–428. https://doi:10.1016/j.jmsy.2022.07.010.

Kaarlela, T., Pitkäaho, T., Pieskä, S., Padrão, P., Bobadilla, L., Tikanmäki, M., Haavisto, T., Blanco Bataller, V., Laivuori, N., and Luimula, M. (2023). Towards metaverse: Utilizing extended reality and digital twins to control robotic systems. *Actuators*, 12(6), Article 2191.

Kaasinen, E., Anttila, A. H., Heikkilä, P., Laarni, J., Koskinen, H., and Väätänen, A. (2022). Smooth and resilient human–machine teamwork as an Industry 5.0 design challenge. *Sustainability*, 14(5), Article 27731.

Kahneman, D., Sibony, O., and Sunstein, C. R. (2021). *Noise. A flaw in human judgment.* William Collins.

Karvonen, A., Kujala, T., Kärkkäinen, T., and Saariluoma, P. (2023). Fundamental concepts of cognitive mimetics. *Cognitive Systems Research*, *82*, 101166.

Kontio, E., Ravyse, W., Saarenpää, T., Pizarro-Lucas, E., Dorado-Diaz, I., Sanchez, P.L., Haavisto,T., and Luimula, M. (2023). Experiences on the creation of a multi-disciplinary course in a metaverse environment, In: Proceedings of the 19th International CDIO Conference (CDIO 2023), Trondheim, Norway, p. 10.

Korteling J. E. (Hans), van de Boer-Visschedijk G. C., Blankendaal R. A. M., Boonekamp R. C., and Eikelboom A. R. (2021). Human versus artificial intelligence. *Frontiers in Artificial Intelligence*, 4, Article: 622364. https://doi:10.3389/frai.2021.622364.

Luimula, M., Haavisto, T., Pham, D. V. D., Markopoulos, P., Aho, J., Markopoulos, E., and Saarinen, J. (2022). The use of metaverse in maritime sector – a combination of social communication, hands on experiencing and digital twins. In AHFE International Conference Proceedings. AHFE Open Access, pp. 123–134. https://doi.org/10.54941/ahfe1001513

MIT Technology Review Insights & Siemens (2023). *The emergent industrial metaverse* (C. Waxer, Author; T. Elsey, Ed.). MIT Technology Review Insights. Retrieved from www.technologyreview.com/2023/03/29/1070355/the-emergent-industrial-metaverse/

Myllylä, M., Karvonen, A., and Koskinen, H. (2024). Design systems for intelligent technology. In: Ahram, T., Karwowski, W., Russo, D., and Di Bucchianico, G. (Eds.) *Intelligent Human Systems Integration (IHSI 2024): Integrating People and Intelligent Systems. AHFE (2024) International Conference. AHFE Open Access, 119.* AHFE International, USA, pp.255–265. https://doi: 10.54941/ahfe1004490

Myllylä, M. T., and Saariluoma, P. (2022). Expertise and becoming conscious of something. *New Ideas in Psychology*, 64, Article: 100916. https://doi:10.1016/j.newideapsych.2021.100916.

Niiniluoto, I. (1993). The aim and structure of applied research. *Erkenntnis*, 38(1), pp.1–21.

Noble, S. M., Mende, M., Grewal, D., and Parasuraman, A. (2022). The Fifth Industrial Revolution: How harmonious human–machine collaboration is triggering a retail and service [r]evolution. *Journal of Retailing*, 98(2), pp.199–208. https://doi.org/10.1016/j.jretai.2022.04.003.

Nokia (2024). Turku University of Applied Sciences creates successful teleoperation solution on Nokia RXRM platform. Nokia. https://onestore.nokia.com/asset/214205

Norros, L., Savioja, P., and Koskinen, H. (2015). Core-task design: A practice-theory approach to human factors. *Synthesis Lectures on Human-Centered Informatics*. Springer.

Oulasvirta, A., Jokinen, J. P. P., and Howes, A. (2022). Computational rationality as a theory of interaction. In: Proceedings of the 2022 CHI Conference on Human Factors in Computing Systems (CHI '22), 359, pp. 1–14. Association for Computing Machinery. https://doi:10.1145/3491102.3517739.

Pinzone, M., Albè, F., Orlandelli, D., Barletta, I., Berlin, C., Johansson, B., and Taisch, M. (2020). A framework for operative and social sustainability functionalities in human-centric cyber-physical production systems. *Computers & Industrial Engineering*, 139, Article: 105132. https://doi.org/10.1016/j.cie.2018.03.028.

Quine, W. V. O. (1980). *From a logical point of view: Nine logico-philosophical essays*. Harvard University Press.

Ripple, W.J., Wolf, C., Gregg, J. W., Rockström, J., Mann, M. E., Oreskes, N., Lenton, T. M., Rahmstorf, S., Newsome, T. M.. Xu, C., Svenning, J.-C., Cardoso Pereira, C., Law, B. E., and Crowther, T. W. (2024). The 2024 state of the climate report: Perilous times on planet Earth. *BioScience*, 74 (12), pp. 812–824. https://doi.org/10.1093/biosci/biae087.

Rožanec, J.M., Novalija, I., Zajec, P., Kenda, K., Ghinani H.T., Suh, S., Veliou, E., Papamartzivanos, D., Giannetsos, T., Menesidou, S.A., Alonso, R., Cauli, N., Meloni, A., Recupero, D.R., Kyriazis, D., Sofianidis, G., Theodoropoulos, S., Fortuna, B., Mladenić, D., and Soldatos, J. (2022). Human-centric artificial intelligence architecture for Industry 5.0 applications. *International Journal of Production Research*, 61(20), pp.6847–6872. https://doi:10.1080/00207543.2022.2138611.

Saariluoma, P., Cañas, J. J., & Leikas, J. (2016). *Designing for Life: A Human Perspective on Technology Development*. Palgrave Macmillan.

Saariluoma, P., Cañas Delgado, J., and Myllylä, M. (2024). Problems of content-based cognitive ergonomics. In: T. Ahram, W. Karwowski, D. Etinger, and T. Mijač (Eds.), *Human Systems Engineering and Design (IHSED2024): Future Trends and Applications*. AHFE (2024) International Conference. AHFE Open Access, 2. AHFE International, USA. http://doi.org/10.54941/ahfe1005545.

Saariluoma, P., and Karvonen, A. (2023). Theory languages in designing artificial intelligence. *AI & Society*, 39, pp. 2249–2258. https://doi:10.1007/s00146-023-01716-y.

Saariluoma, P., Karvonen, A., and Sorsamäki, L. (2021). Human digital twins in acquiring information about human mental processes for cognitive mimetics. In: Tropmann-Frick, M., Jaakkola, H., Thalheim, B., Kiyoki, Y., and Yoshida, N. (Eds.) *Information Modelling and Knowledge Bases XXXIII (EJC 2021). Frontiers in Artificial Intelligence and Applications, 343*. IOS Press, pp.163–176. https://doi: 10.3233/faia210484.

Saariluoma, P., Myllylä, M., Karvonen, A., Luimula, M., and Aho, J. (2024). A human digital twin for the M-machine. *Discover Artificial Intelligence*, 4(1), 61.

Schön, D. A. (2017). *The reflective practitioner: How professionals think in action*. Routledge.

United Nations (2015). *Transforming our world: the 2030 Agenda for Sustainable Development*. https://sdgs.un.org/2030agenda (Accessed 11 October 2024).

Vacchi M, Siligardi C, Demaria F, Cedillo-González EI, González-Sánchez R, and Settembre-Blundo D. (2021). Technological sustainability or sustainable technology?

A multidimensional vision of sustainability in manufacturing. *Sustainability*, 13(17), 9942. https://doi.org/10.3390/su13179942.

Vinuesa, R., Azizpour, H., Leite, I. Balaam, M., Dignum, V., Domisch, S., Felländer, A., Langhans, S. D., Tegmark, M., and Nerini, F. F. (2020). The role of artificial intelligence in achieving the Sustainable Development Goals. *Nature Communications*, 11, Article: 233. https://doi.org/10.1038/s41467-019-14108-y.

Wahlström, M., Tammentie, B., Salonen, T. T., and Karvonen, A. (2024). AI and the transformation of industrial work: Hybrid intelligence vs double-black box effect. *Applied Ergonomics*, *118*, 104271.

Xu, X., Lu, Y., Vogel-Heuser, B., Wang, L. (2021). Industry 4.0 and Industry 5.0—Inception, conception and perception. *Journal of Manufacturing Systems*, 61, pp.530–535. https://doi.org/10.1016/j.jmsy.2021.10.006.

8 Design and Production of Large-Scale Products by Robotic Extrusion-Based Additive Manufacturing

Artur Costa, Afonso Santos, Álvaro M. Sampaio, and António J. Pontes

INTRODUCTION

This chapter explores the emergence of product design and development for urban spaces using robotic extrusion, examining the benefits and influence of additive manufacturing (AM) processes within this area of application. The constant increase in urban areas has distanced people from nature, and this shows major implications for health. Green areas within urban spaces provide spaces for leisure and recreation, which leads to social improvements. Herein displays a process methodology that seeks to create products that improve social interaction within urban green spaces focusing on leisure. To achieve this, sustainable products will be developed through a large-scale robotic extrusion-based system, combining the versatility and design possibilities that AM allows with the associated sustainability of recycled plastic waste.

The expansion of urban environments and the rising number of residents in these areas have contributed to a greater separation between people and nature, which poses health risks (WHO, 2017). Urban green spaces are crucial for enhancing the quality of life for both inhabitants and pedestrians, offering advantages not only for the environment but also for both individual and public health (Hartig et al., 2014; WHO, 2016). Given the separation between urban dwellers and nature, it is vital to establish new green spaces or preserve those that already exist (WHO, 2017). This presents a chance to further develop outdoor areas conducive to socializing and studying, fostering social connections in a more natural and engaging setting. The connection between urban living and green spaces is tied to their availability and quality (Hartig,

DOI: 10.1201/9781003625360-8

2007; Lachowycz & Jones, 2013; Yang et al., 2016), thus creating an opportunity for innovation.

Robotic extrusion-based additive manufacturing (REBAM) systems enable the production of significantly larger structures compared to traditional 3D printing technologies (Choo et al., 2019). Among AM techniques, material extrusion (ME) has become one of the most prevalent methods for producing polymer components (Chesser et al., 2019). In this context, ME is often referred to as fused filament fabrication (FFF), where polymer filament is used as a feedstock (Go et al., 2017).

Nevertheless, the use of polymer filament in FFF systems has its drawbacks, particularly regarding production speeds. Most commercially available 3D printers are limited by small build volumes and slow material deposition rates, primarily due to the constraints of filament-based extrusion systems (Chesser et al., 2019; Go et al., 2017). In contrast, REBAM addresses these challenges by utilizing polymer pellets as the raw material, which enables much faster deposition rates (Chesser et al., 2019). However, while these increased build speeds are advantageous, they complicate the achievement of the desired geometric accuracy and surface quality, especially when larger layer sizes are employed, resulting in diminished part resolution and an increase in defects. These compromises require the creation of new design principles and printing parameters tailored for REBAM systems (Roschli et al., 2019).

Grasping the constraints and benefits of this process, as well as how different materials behave, is crucial for enhancing production efficiency and speed. Based on these assessments, this chapter delves into the design constraints posed by materials by comparing two distinct types: a recycled polypropylene mixed with 30% calcium carbonate (PP/$CaCO_3$) and a general-purpose virgin acrylonitrile butadiene styrene (ABS).

IMPORTANCE OF URBAN SPACES AND THE RELATIONSHIP WITH LEISURE

Urban spaces are an integral part of the cultural, political, and economic life of society, from the earliest civilizations to the present day. Historically, the form, function, and use of urban spaces have varied dramatically, based on cultural context, yet preserving a set of similar characteristics (Stanley et al., 2012).

The importance of urban spaces is recognized by different authors, attributing to them social and political value, while also benefitting the health (Stanley et al., 2012; Hartig et al., 2014; Ulrich, 1983; WHO, 2016) of communities and their residents. The literature on the importance of leisure in urban spaces is currently extensive, as is their importance for populations, and it has been a topic of study for several decades.

Urbanization is driving a steady increase in the number of people residing in cities. In Europe, projections indicate that by 2020, approximately 75% of the population will inhabit urban areas. Living in urban areas often restricts access to green spaces and can expose residents to various environmental hazards, including pollution, scarcity of natural resources, and the effects of climate change. Addressing these issues is crucial for ensuring that cities can offer healthy and sustainable environments for their inhabitants (WHO, 2017).

The relationship between health and access to green spaces is complex and can be viewed in different ways depending on the population under study and the characteristics of the location. Several models have been proposed to explain the relationship between health and green spaces. Hartig et al. (2014) suggested four main and linked ways in which nature and green spaces can contribute to health: improved air quality, physical activity, reduced stress, and increased social contact.

Green spaces play an important role in fostering social relationships and promoting a sense of community. Lachowycz & Jones (2013) highlighted physical activity, relationship with nature and rest, and social activities and interactions as the biggest factors in improving health in relation to urban green spaces. The increased appeal of such environments can encourage people to spend more time outdoors and help stimulate physical activity.

Taking all of this into account, the use of REBAM for large-scale urban products such as benches and other street furniture demonstrates its versatility and scalability, making it a promising solution for creating large, complex, and sustainable structures that meet the aesthetic and functional demands of modern cities. The use of robotic equipment in AM leads to new production paradigms, emphasizing three critical factors in the production process: the manufacturing process, the material used, and the part geometry to be produced. Understanding these three factors and their respective constraints leads to better quality and efficiency in part production.

METHODS

Robotic Equipment and Process Methodology

Today, a wide variety of robots can be found on the market. These machines were initially developed to perform tasks on production lines, and most are still used for this purpose today. That said, robots can vary according to the task they perform, the load, or the range they need to reach. The anthropomorphic robotic arm is a multipurpose robot that is easily adapted to any task and is therefore used in all types of industry (Sampaio et al., 2023; Lopes et al., 2023). Within this type of robot, there are different models suitable for different functions. Brands such as KUKA and ABB stand out in the production of these instruments.

The main equipment used in this study includes a KUKA KR120 R3100-2 6-axis robotic arm, a single-screw polymer pellet extruder with four heating zones, and a heated aluminium bed that can reach temperatures of up to 140°C, providing process flexibility and precision in printed layer control. The trajectories of the robotic arm were programmed using Grasshopper software, which serves as the parametric environment within Rhinoceros® modelling software. The KUKA|prc (parametric robot control) add-on was used to virtually simulate the movements of the robotic arm and to export the printing code in KRL (KUKA Robot Language) format. Grasshopper allows for complete control over the generated toolpaths, which consequently determines the movements of the robotic arm. For this reason, Grasshopper is frequently applied in the field of robotic programming for AM (Serrano & Bares, 2023) and even for FFF printer programming (Cuevas & Pugliese, 2020).

Nevertheless, the material used is the primary factor that restricts the geometry of the produced parts. As with conventional 3D printing, but on a larger scale, this process presents a new set of restrictions in the production of parts and in the production parameters, while maintaining the restrictions found in FFF processes (Roschli et al., 2019). Among the limitations and precautions to be taken in the design are the following:

- Over-extrusion: The issue of excessive material being extruded from the nozzle onto the build platform is a common challenge. This problem stems from the flow rate parameter and can result in several complications, including oozing, blobs, stringing, and a decline in overall print quality.
- Delamination: This occurs when the printed layers separate from each other or from the printing base. This phenomenon is caused by the rapid cooling of the material. In REBAM, as the parts produced are larger, this leads to a longer production time for each layer, which results in a greater thermal gradient.
- Surface finish: The REBAM process is carried out layer by layer using a large polymer extruder. This results in larger printed layers, which makes them more noticeable. The surface finish can be improved if the printed surface is sanded or machined.
- Overhangs: As it is not feasible to produce support material in these technologies, it is important to consider the maximum printing angle. Exceeding this angle can result in the layers folding over and collapsing, which leads to production failure.

The development of large-scale products produced using robotic extrusion allowed for the study of the feasibility of this technology, while avoiding overstepping its limitations. To achieve this, the initial step involved creating two sample components with distinct attributes to evaluate the maximum printing angle (overhang) and the range of curvature radii that the product could accommodate. This ensured proper printing and served as a proof of concept.

Preparing a Program File – From a 3D Model to a KRL Code

In this chapter, an explanation is presented on how programming is carried out using Grasshopper to generate the robot's toolpath, taking advantage of the KUKA robotic arm's six degrees of freedom to create customized deposition strategies.

To begin with, it's essential to grasp how a computer-aided design (CAD) model is transformed into code that the robot controller can interpret. Then, for the visual programming aspect, Grasshopper was utilized by incorporating the KUKA|prc add-on, as discussed earlier, to be able to simulate the robot's movements in a virtual setting and generate the necessary code for printing. A Grasshopper script was created to convert a given geometry into a continuous trajectory that the robot's end effector (in this case, the extruder nozzle) Tool Control Point (TCP) needs to follow. Basic components in Grasshopper were employed to create the toolpath, converting it into a sequence of coordinates that will dictate the robot's actions. Figure 8.1

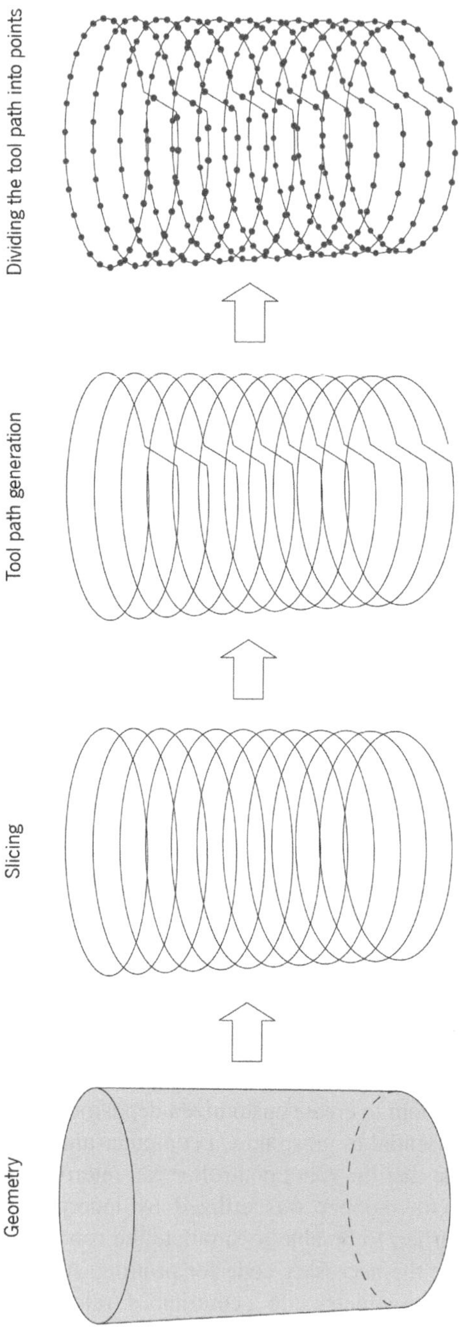

FIGURE 8.1 Programming workflow for transforming a geometry into Kukalprc commands.

shows a schematic of how this program works. To segment the geometry, the *contour* command was utilized. Following this, a series of command groups, or *clusters*, were used to develop the continuous path. This path is then broken down into individual points, which generate planes that encompass all the critical data required for the robot's movement.

The alignment of these planes determines the orientation of the tool. After establishing the planes, they serve as the input commands for KUKA|prc. This plugin allows us to set the initial and final positions of the robot's end effector, along with adjusting the robot's speed (Cuevas & Pugliese, 2020).

Regardless of the programming method used, the goal is to obtain a code file, in textual language, that is compatible and can be interpreted by the control system. Since the operated programming is focused on the KUKA language, the generated code is exported as KRL in a .src file. This file contains the information needed for the robot to execute its movements (initial and final position, linear speed, etc.), as well as information for external elements such as mobile bases or the tool.

Despite Grasshopper's versatility, during programming, various constraints may arise in both the development and production stages of a component. Many of these issues stem from hardware limitations, such as poor communication between the robot and the extruder. Without the robot's controller managing the extruder, there is a lack of dynamic interaction between the robot's movements and the material being extruded. This limitation hinders the ability to create specific structural variations in the printed item. Consequently, the design is restricted by the programming and hardware capabilities of the robotic system, resulting in a component that can only maintain a continuous profile. While producing infills is feasible, it requires careful definition of the toolpath based on the part's geometry to ensure accurate generation. Given these limitations, creating fully enclosed printed components proves difficult, allowing only for open structures at both the top and bottom.

PRINTING PARAMETERS AND DESIGN GUIDELINES

As previously mentioned, it is crucial to be aware of the software and programming restrictions during the design phase. Additionally, there are physical design limitations to consider. In this research, the primary constraint under investigation was the overhang angle.

In this context, the overhang angle mentioned refers to how much a layer's outline deviates from the vertical build direction in relation to a horizontal surface. This important overhang angle acts as a fundamental factor in assessing whether a component is suitable for AM (Roschli et al., 2019; Serrano & Bares, 2023). For example, a straight cylinder has an overhang angle of 0°, while a cylinder with a capped top has an overhang angle of 90°. This approach to measuring angles allows for more precise simulations of CAD models using draft angle analysis software. The maximum overhang angle is a significant design factor in REBAM, as it's not feasible to produce support structures with the available resources.

To determine the maximum overhang angle for each material, test parts were fabricated using the parameters listed in Table 8.1. The test part model was designed

TABLE 8.1
Printing parameters for the overhang angle's test parts

	$PP/CaCO_3$	ABS
Layer height	3.0 mm	3.5 mm
Robot speed	8 mm/s	8 mm/s
Extruder's screw speed	3 RPM	5 RPM
Extruder's screw speed (for 1st layer)	10 RPM	10 RPM
Printing temperatures		
Zone 1	160 °C	180 °C
Zone 2	170 °C	200 °C
Zone 3	180 °C	205 °C
Zone 4	185 °C	210 °C

having a cylindrical base, as its circular design improves adhesion to the printing surface and reduces warping. Sections with gradually increasing angles were printed along the build direction, beginning at 0° and followed by increments of 10°, 20°, 30°, 35°, 40°, 45°, 50°, and 55°.

It is important to mention that different materials require different production parameters. As was observed during previous layer deposition testing, with a fixed nozzle diameter of 4 millimetres, the printed layer height for a virgin material reached 3.5 millimetres, while for composite materials was 3.0 millimetres. Considering that the test piece was produced identically for both materials, the robot's printing speed was kept constant. However, it was necessary to adjust the rotation speed of the extruder screw according to the material being processed. Since both have different viscosity values, that means that different amounts of molten material are extruded using the same screw speed (Gupta & Taufik, 2021). To ensure the first layer's adhesion to the print bed, a higher rotation speed was applied for that initial layer. Temperature settings were also adjusted based on the material used.

RESULTS AND DISCUSSION

Test Parts

Figure 8.2 illustrates the results of the overhang angle test using a guideline as reference. For $PP/CaCO_3$ (1), this material reached a maximum overhang angle of 40°, beyond which delamination occurred due to insufficient layer self-support. Conversely, virgin ABS (2) reached an overhang angle of 50°. However, at this angle, a deviation of the part from the guideline was observed. Therefore, it is concluded that the maximum overhang angle for this material is 45°. This is consistent with the fact that features with an angle of inclination greater than 45° will typically need to be held by additional support material (Roschli et al., 2019; Isa et al., 2018).

The results of these tests are shown in Table 8.2, together with a measurement of the wall thickness of the produced parts.

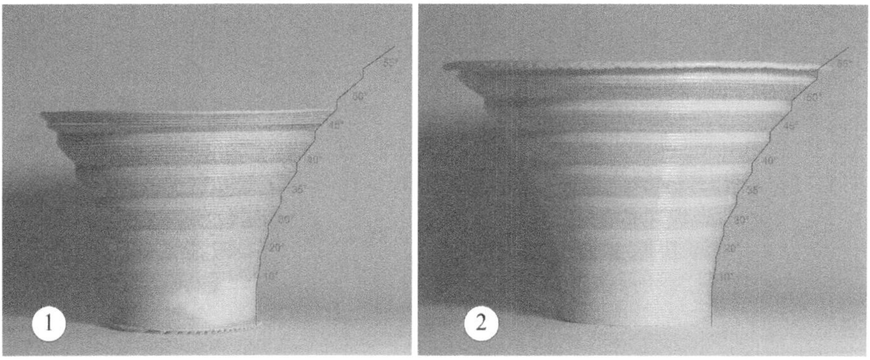

FIGURE 8.2 Maximum overhang angle test for (1) PP + 30% $CaCO_3$ and (2) ABS. Photograph by the author.

TABLE 8.2
Overhang angle test results

	PP/$CaCO_3$	ABS
Wall thickness	8 mm	10 mm
Maximum overhang angle	40°	45°

In the upcoming section of the chapter, an assessment is conducted on a test part focused on determining the minimum curvature of a closed corner, giving a sense of how tightly one can print curved layers without superimposing the deposited material.

To ensure smooth and uninterrupted motion along toolpaths with sharp corners, most computer numerical control (CNC) systems utilize a local smoothing method. This approach employs a predefined curve or spline to soften the corners, allowing the system to slow down and adjust the feed direction according to the machine's kinematics. When sharp corners are closely spaced, the path speed is reduced significantly to improve accuracy (Tajima & Sencer, 2017). This concept also applies to robotic arms. Given that the extrusion rate remains constant for each layer that is printed and that, since up until this stage, there had been no control over the extrusion rate or travel speed of the robot arm during a single layer, the material tends to accumulate in sharp corners due to over-extrusion. In Figure 8.3, we can assess the minimum curvature radius of a corner that can be printed without overlapping material.

One side of the part features three consecutive triangles that converge to form a closed shape resembling a rectangle. The intersections of these triangles create five corners, each with a 30° angle. Different curvature radii were assigned to these vertices, with the first vertex having no curvature. The subsequent vertices have curvature radii of 5 mm, 10 mm, 15 mm, and 20 mm, moving from left to right. The corners with radii of 15 mm and 20 mm show fewer imperfections originated from

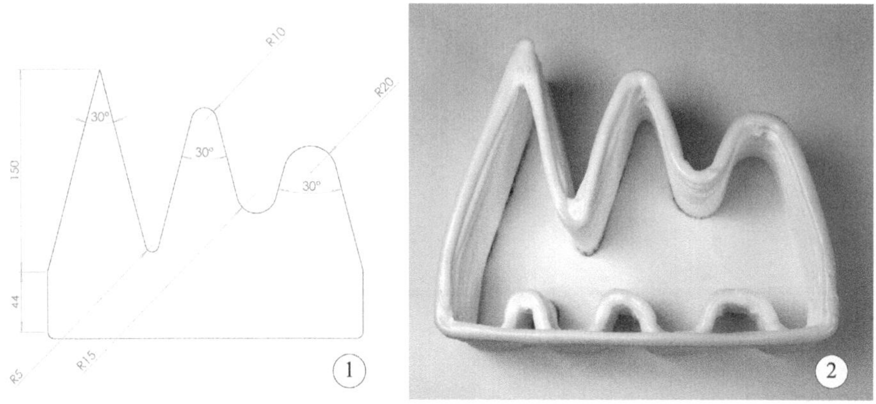

FIGURE 8.3 Top view of the second test part (1) with the dimensions of the corner angles and their respective curvature radius and (2) with the dimensions in millimetres (mm). Photograph by the author.

material accumulation by comparison to the first three corners, which exhibit significant overlap.

The results obtained from this test part are analogous to each one of the materials, since they are only dependent on the equipment constraints of the printing process itself and serve as a benchmark for adequate overall surface finish of a final product.

To examine the behaviour of the materials under different geometrical configurations and identical conditions, the parts shown in Figure 8.4 were manufactured. The objective of this test was to validate the overhang angle that was previously tested, but in a part with increased design complexity that differs from a cylinder profile and having rounded edges with fillet radii between 15 and 20 millimetres as an established geometric limitation. Additionally, the design incorporates negative and positive angles, with the maximum angle reaching 50°, to evaluate the behaviour of the material at this angle under different conditions. As the part is symmetric about the central plane, this allows for a direct comparison between negative and positive angles.

The printing parameters applied to produce both parts are presented in Table 8.3. As done previously, the printing speed was maintained constant, while the extruder's screw speed was adjusted. The layer height was adjusted according to the type of material, resulting in varying printing times.

In addition to the first test part produced to determine the maximum overhang angle, the subsequent test part, designed to function as a bench, provides further insights on this matter. Since this part was intentionally designed having 50° angles, the relevant data are mainly drawn from these geometric features. Figure 8.5 highlights this information, particularly on one of the sides where the angle is positive.

The PP composite material has yielded the worst results due to its higher shrinkage rate and lower printable overhang angle. As shown on the left side of the figure, PP/$CaCO_3$ (1) did not exhibit delamination, although being a filled thermoplastic composite that is predisposed to this phenomenon; however, notable layer sagging was

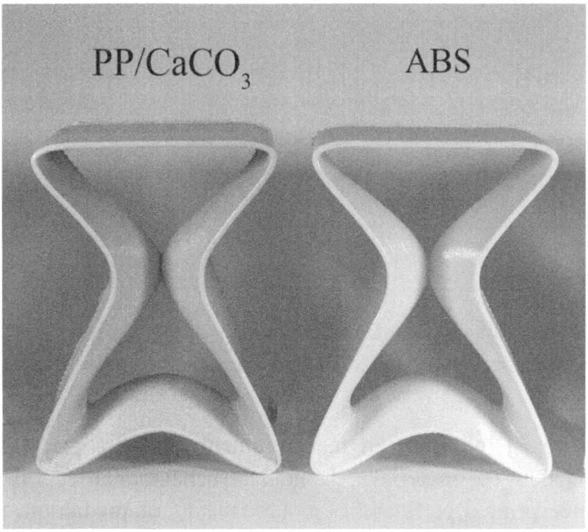

FIGURE 8.4 Test parts produced with both materials. Photograph by the author.

TABLE 8.3
Printing parameters and specifications for each produced part

	$PP/CaCO_3$	ABS
Layer height	3.0 mm	3.5 mm
Layers	104	89
Robot speed	16 mm/s	16 mm/s
Extruder's fuse rotation	9 RPM	9 RPM
Extruder's fuse rotation (for 1st layer)	25 RPM	18 RPM
Printing time	2.90 h	2.50 h
Printing temperatures		
Zone 1	160 °C	180 °C
Zone 2	170 °C	200 °C
Zone 3	180 °C	205 °C
Zone 4	185 °C	210 °C

evident in the specified region. The degree of sagging is a critical factor in determining the stability of the printed structure, as it affects the integrity of adjacent layers. The results obtained for ABS (2) were found to be consistent with the initial overhang angle test. While some sagging was observable with this material, it did not significantly affect the stability of the printed part.

Regarding the negative angles, which are oriented in a direction opposite to those illustrated in Figure 8.5, they exhibited less sagging than the positive angles and did not present significant challenges with respect to the stability or production of the part.

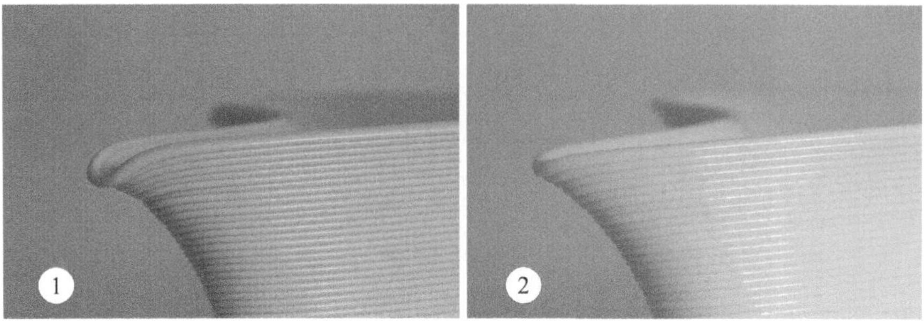

FIGURE 8.5 Material sagging in one of the corners of the test part at a 50° overhang for (1) $PP/CaCO_3$ and (2) virgin ABS. Photograph by the author.

The results registered from all the aforementioned test parts presented valuable insight on how the material's rheological behaviour affects the geometry of intermediate-size printed parts and is a stark means of prediction, using minimal wasted material, to simulate the outcome of definite REBAM products using the same grade of materials.

Concept Design and Product Development

After establishing the design criteria for producing part components through this AM method, final products were conceptualized and created to explore the maximum dimensional capabilities of the technology. To produce the REBAM products, the same virgin and filled thermoplastic materials that were studied earlier were employed for this purpose by programming, for each of them, specified toolpath strategies and applying distinct process parameter inputs.

In order to develop a project that would show the potential of REBAM in urban spaces, the outdoor spaces of the Azurém Campus of the University of Minho were studied. The integration of additive systems of this magnitude into such a busy public area allows for its dynamization in a way that focuses on innovation (Tedeschi, 2014). To satisfy user needs, it is first necessary to identify them so then one may understand how to solve them. For users of the venue, the lack of seating space is obvious. In busy situations, the sheltered space may prove insufficient. In this sense, there are several opportunities to improve the site, including the installation of benches or other similar structures that encourage socializing, study, and leisure activities. Not to mention the fact that the site has a small number of fixed concrete benches that are not comfortable and are concentrated in a specific area rather than distributed around the campus.

The concept design of the proposed REBAM product takes advantage of the concrete benches already in place and aims to create an extension of them, adding comfort and functionality to the existing ones. Once the concept to be applied to the space had been defined, its shape was worked out with the inclusion of backrests. At this stage of development, Rhinoceros software was used to create different organic form iterations, considering the real dimensions of the parts and applied ergonomics

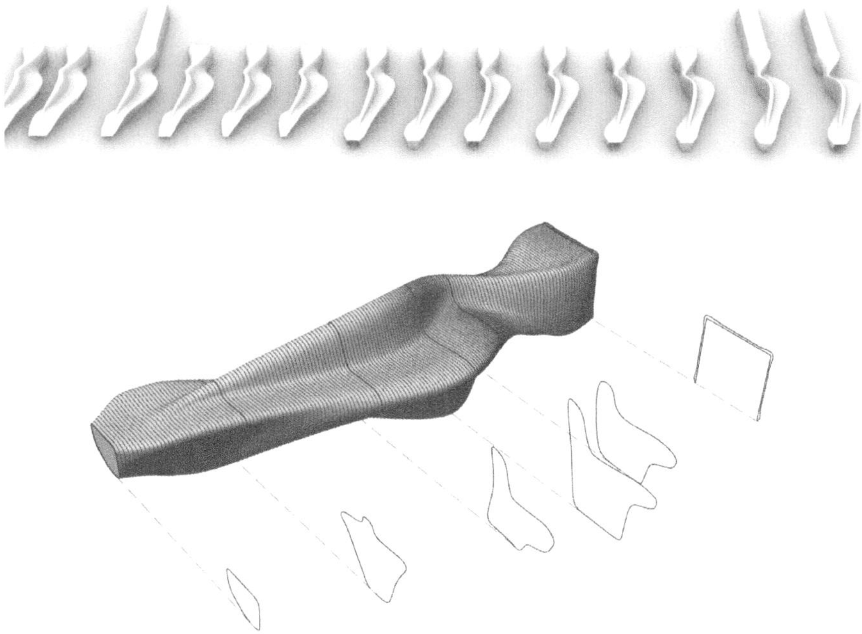

FIGURE 8.6 Concept iterations for the concrete bench extension (top); sections for modelling the geometry under analysis (bottom).

(Figure 8.6). The selected concept design shown among the illustrations below was produced in a vertical position, starting at the square-shaped end.

Before the production stage, it was necessary to create a program in the Grasshopper software that would allow the parts to be transformed into a continuous movement of the robotic arm. The aim of creating a program is to facilitate the transformation of any geometry, even if it is external to the Grasshopper software, into a movement code that can be read by the robot. Using Grasshopper, the intent was to create a parametric code that allowed important parameters to be changed for printing parts. These parameters include the layer height, the orientation and positioning of the geometry in space, and the deposition speed of the robotic arm.

Figure 8.7 shows the draft angle analysis regarding the geometry areas that are closer to the limit of process feasibility. Subsequently, Figure 8.8 presents an overview of the product's print code, programmed with the Kukalprc add-on, having the ability to change process parameters such as build position and orientation, robot deposition speed, layer height, and other parametric functions according to need.

Regarding PP/$CaCO_3$, due to its infilled nature, it was necessary to create design constraints to mitigate, as much as possible in uncontrolled room temperature conditions, the residual tensions created by the rapid crystallization of the thermoplastic matrix of this composite material. These molecular forces are responsible for creating warpage, so precautions were taken to distribute heat through various points

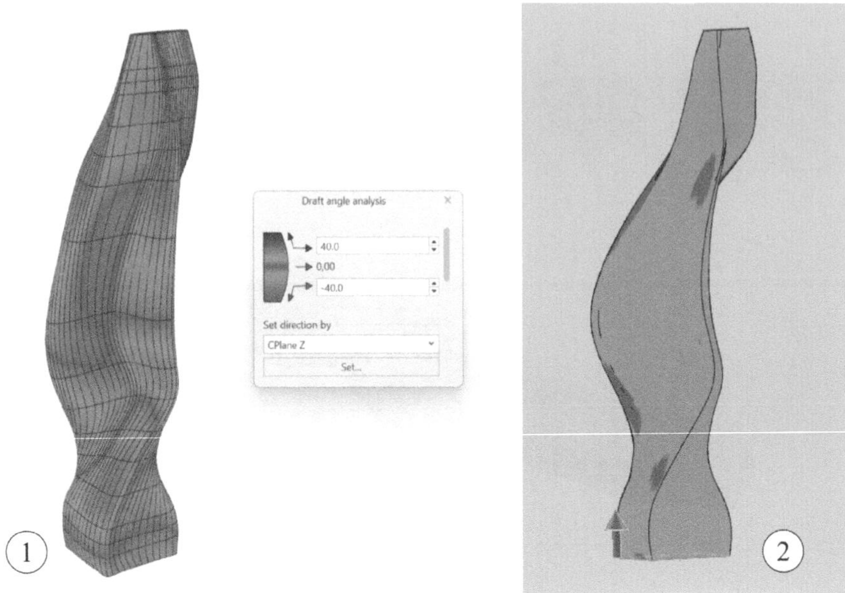

FIGURE 8.7 Draft angle analysis of proposed CAD geometry models using Rhinoceros (1) and SolidWorks (2) software.

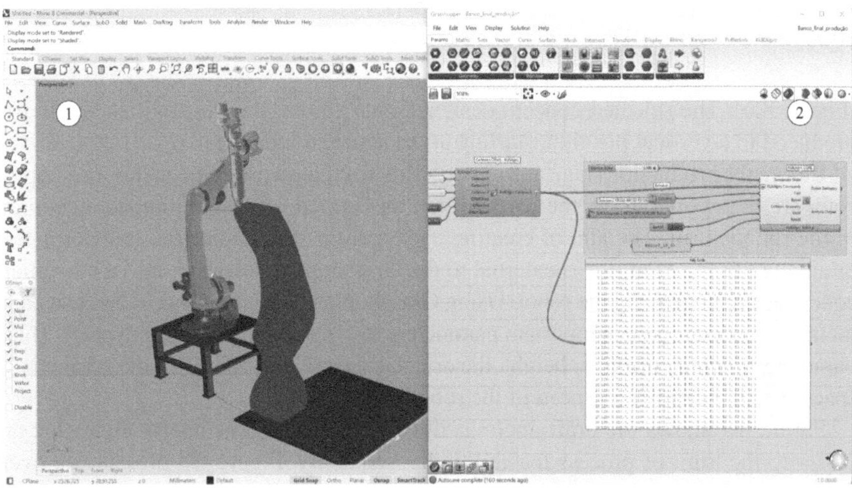

FIGURE 8.8 Display of one the large-scale product's print codes, programmed with the Kukalprc add-on. The Rhinoceros environment (1) shows the production simulation, and the Grasshopper environment (2) shows some KUKAlprc commands together with a KRL text.

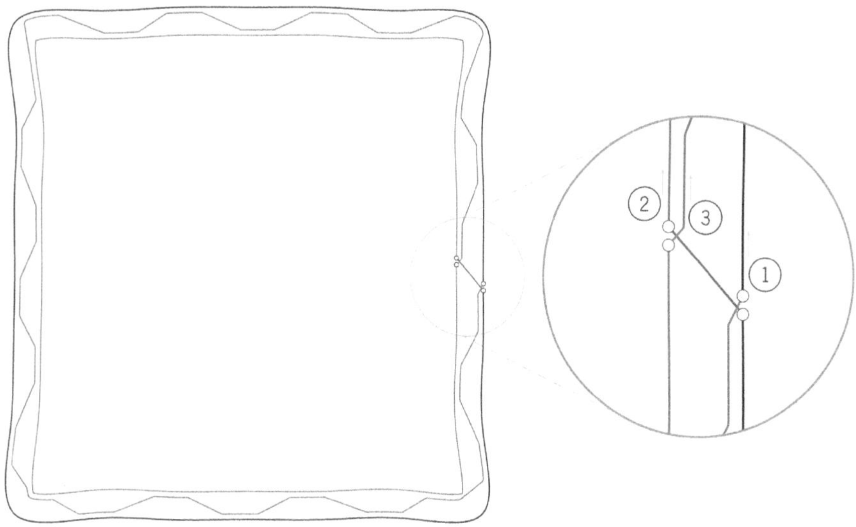

FIGURE 8.9 Illustration of the infill toolpath to produce the internal truss, highlighting the deposition order of the outer contour (1), inner contour (2), and infill contour (3).

of contact between different contours within the same layers, thereby alleviating said tension along the build orientation (see Figure 8.9).

As can be seen in Figure 8.10, although this method did prove to be overall effective in controlling the repercussions of warpage, it did, however, play a role in reducing the adhesion strength between subsequent layers. This, in turn, increased the propensity to generate horizontal cracks (delamination) along some angled surface zones, since vertical tensions located in these reduced contact areas were now stronger than the more stabilized horizontal residual tensions.

Figure 8.11 illustrates four distinct phases of the total production process. The first phase (1) denotes the initial hour of production, where ensuring proper adhesion of the first layers to the printing base is essential. The second (2) and third (3) phases indicate a midpoint in production, during which adjustments to printing parameters are made after each stage. The final phase (4) represents the completion of the process, culminating in the finished part that must be adequately removed from the printing base for the further instalment in an urban environment. In the following image (Figure 8.12), it is possible to observe the product on-site serving the purpose for which it was created, functioning as an extension of one of the fixed benches among the university campus.

Similarly, Figure 8.13 and Figure 8.14 exhibit, respectively, the final production stage and outcome of the second developed bench produced with virgin ABS.

Both large-scale products were produced with the respective layer heights and extrusion temperature profiles according to the values that were referenced earlier in this chapter. However, due to the added geometric complexity and increased variability of the layer perimeter, different robot printing speeds and screw speeds had to be set for various sections of the two products, in compliance with how the rheological behaviour of each material affects their flow rate. For this reason, different total printing times were reached.

FIGURE 8.10 Initial production stage of the bench, exhibiting the internal truss disposition along the build orientation; example of generated cracks, highlighted by the red arrow (on the right). Photograph by the author.

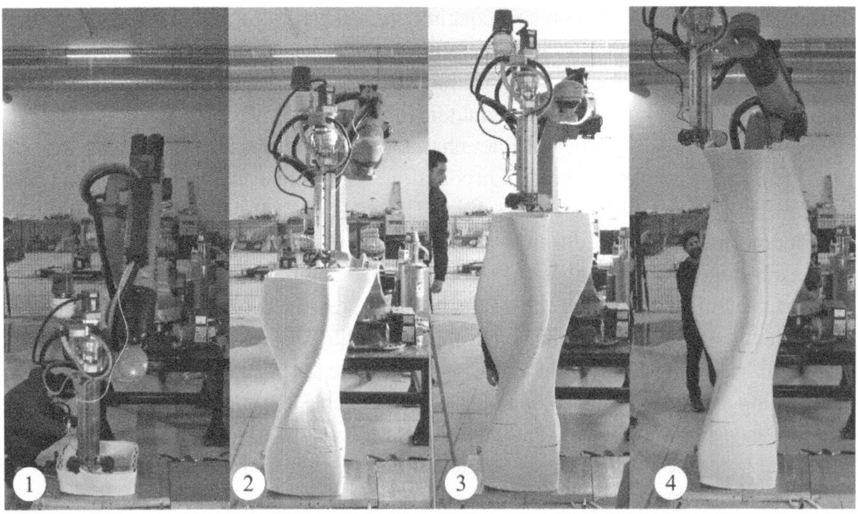

FIGURE 8.11 Different production phases of the developed bench shape using PP/$CaCO_3$. Photograph by the author.

FIGURE 8.12 Large-scale PP/$CaCO_3$ product installed in the designated urban green space site (first bench model). Photograph by the author.

FIGURE 8.13 Final production stage of the second developed bench shape using virgin ABS. Photograph by the author.

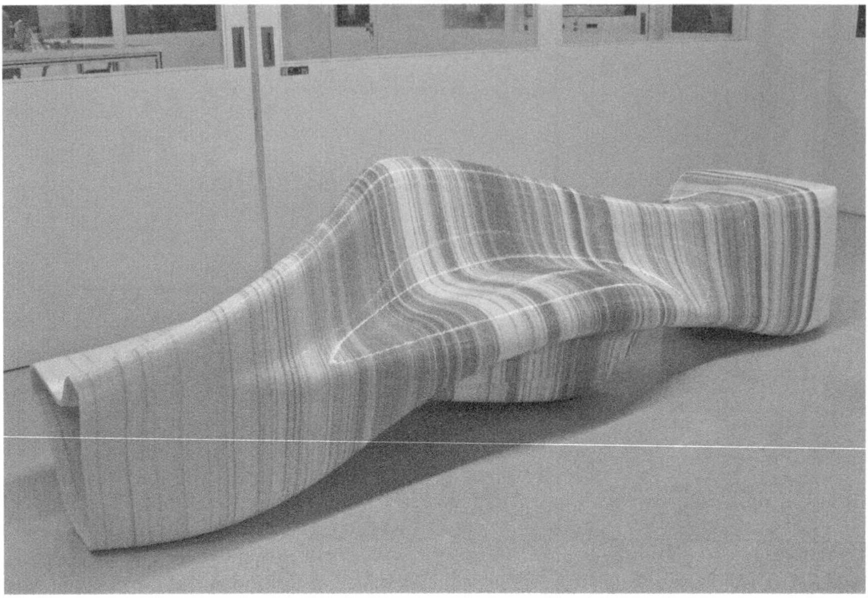

FIGURE 8.14 Large-scale virgin ABS product ready to be placed within an urban space setting (second bench model). Photograph by the author.

TABLE 8.4
Process results for each large-scale product

	$PP/CaCO_3$	ABS
Product height	2.70 m	2.70 m
Layers	900	772
Printing time	26.30 h	21 h
Processed material	100 Kg	75 Kg
Waste material	<0,1 Kg	~0 Kg

The main process results of these large-scale products are summarized in Table 8.4. The employed manufacturing method demonstrated significant advantages, particularly in terms of design flexibility (even with the noted limitations) and reduced material consumption when compared to alternative methods for large components. This makes it a more sustainable and environmentally friendly option.

Ultimately, the design of these environmentally friendly models showcased that this technology can be customized to utilize only the essential amount of plastic needed to ensure structural integrity for public use. Additionally, the equipment and products are characterized by their low energy consumption, eliminating the need for excessive energy expenditure to enhance the manufacturing process—both in terms of material usage (minimal energy required for raw material melting) and the electric machinery involved.

CONCLUSIONS

The creation of large-scale products, specifically large customizable benches for urban spaces, was achieved through an advanced REBAM technology that utilizes ME with a robotic arm. This cutting-edge and transformative product design aims to enhance social interactions while also enriching green spaces in urban environments, thanks to the sustainability provided by its recycled plastic waste materials and optimum usage of feedstock.

It was determined that calcium carbonate-filled recycled PP, being readily available, was the most appropriate material for the case study, as it offers further adequate structural robustness and mechanical durability, along with a notable ability to endure repeated human interaction. Nevertheless, more conventional virgin materials, such as the additionally studied ABS, proved to be more workable in terms of geometric liberties, primarily due to the improbable occurrence of delamination. A methodology was, therefore, first established to identify the design constraints and processing limitations for the production stages with these two materials. The creation of standard and intermediate-size test parts with specific feature analysis facilitated the establishment of a maximum printing angle of 40º, with the minimum curvature radius needing to be 15 millimetres or greater for PP/$CaCO_3$. ABS demonstrates a superior printing angle, reaching 45°.

The research presented in this chapter led to the creation of large-scale, innovative, and elegantly designed items, in contrast to existing products that accomplish similar functions, measuring as large as 2.70 metres in width. Their intended use is as outdoor fixtures to be set in urban areas, offering an alternative means of facilitating social interaction.

REFERENCES

Chesser, P., Post, B., Roschli, A., Carnal, C., Lind, R., Borish, M., & Love, L. (2019). Extrusion control for high quality printing on Big Area Additive Manufacturing (BAAM) systems. *Additive Manufacturing*, 28, 445–455. https://doi.org/10.1016/j.addma.2019.05.020

Choo, K., Friedrich, B., Daugherty, T., Schmidt, A., Patterson, C., Abraham, M. A., Conner, B., Rogers, K., Cortes, P., & MacDonald, E. (2019). Heat retention modeling of large area additive manufacturing. *Additive Manufacturing*, 28, 325–332. https://doi.org/10.1016/j.addma.2019.04.014

Cuevas, D., & Pugliese, G. (2020). *Advanced 3D printing with Grasshopper Clay and FDM*. North Haven.

Go, J., Schiffres, S. N., Stevens, A. G., & Hart, A. J. (2017). Rate limits of additive manufacturing by fused filament fabrication and guidelines for high-throughput system design. *Additive Manufacturing*, *16*, 1–11. https://doi.org/10.1016/j.addma.2017.03.007

Gupta, A. K., & Taufik, M. (2021). Effect of process variables on performances measured in filament and pellet based extrusion process. *Materials Today: Proceedings*, 47, 5177–5184. https://doi.org/10.1016/j.matpr.2021.05.508

Hartig, T. (2007). Three steps to understanding restorative environments as health resources. In *Open Space: People Space*. Taylor & Francis.

Hartig, T., Mitchell, R., De Vries, S., & Frumkin, H. (2014). Nature and health. *Annual Review of Public Health*, 35(1), pp. 207–228.

Isa, M. A., Yiğit, I. E., & Lazoglu, İ. (2018). Analysis of Build Direction in Deposition-Based Additive Manufacturing of Overhang Structures, In Proceedings of the 29th Annual

International Solid Freeform Fabrication Symposium. University of Texas, Austin. http://dx.doi.org/10.26153/tsw/42318

Lachowycz, K., & Jones, A. P. (2013). Towards a better understanding of the relationship between greenspace and health: Development of a theoretical framework. *Landscape and Urban Planning*, 118, 62–69. https://doi.org/10.1016/j.landurbplan.2012.10.012

Lopes, A.C., Sampaio, A.M., Fernandes, F., Vaz, P.M., & Pontes, A.J. (2023). Design and industrial implementation of a multi-functional part produced by powder bed fusion. *Progress in Additive Manufacturing*, 9, pp. 935–945. https://doi.org/10.1007/s40964-023-00493-w

Roschli, A., Gaul, K. T., Boulger, A. M., Post, B. K., Chesser, P. C., Love, L. J., Blue, F., & Borish, M. (2019). Designing for big area additive manufacturing. *Additive Manufacturing*, 25, pp. 275–285.

Sampaio, A.M., Lima, A., Silva, C., Carneiro, V.H., Fernandes, F., Fernandes, A. & Pontes, A.J. (2023). Design and development of an Automatic Optical Inspection (AOI) system support based on digital manufacturing. *Procedia CIRP*, 119, 15–20. https://doi.org/10.1016/j.procir.2023.03.080

Serrano, A. N., & Bares, J. M. E. (2023). Advances in large-scale robotic 3D printing with plastic pellets. In *IntechOpen eBooks*. https://doi.org/10.5772/intechopen.109438

Stanley, B., Stark, B., Johnston, K., & Smith, M. (2012). Urban open spaces in historical perspective: A transdisciplinary typology and analysis. In *Urban Geography*, 33(8), pp. 1089–1117. https://doi.org/10.2747/0272-3638.33.8.1089

Tajima, S., & Sencer, B. (2017). Global tool-path smoothing for CNC machine tools with uninterrupted acceleration. *International Journal of Machine Tools & Manufacture*, 121, pp. 81–95.

Tedeschi, A. (2014). *AAD Algorithms-Aides Design: Parametric Strategies using Grasshopper*. Le Penseur.

Ulrich, R. S. (1983). Aesthetic and affective response to natural environment. In *Behavior and the Natural Environment* (pp. 85–125). Springer US. https://doi.org/10.1007/978-1-4613-3539-9_4

World Health Organization (WHO) (2016). Urban green spaces and health (No. WHO/EURO: 2016-3352-43111-60341). World Health Organization, Regional Office for Europe: Copenhagen, Denmark.

World Health Organization (WHO) (2017). Urban green spaces: a brief for action. World Health Organization, Regional Office for Europe: Copenhagen, Denmark.

Yang, Y. C., Boen, C., Gerken, K., Li, T., Schorpp, K., & Harris, K. M. (2016). Social relationships and physiological determinants of longevity across the human life span. *Proceedings of the National Academy of Sciences of the United States of America*, 113(3), 578–583. https://doi.org/10.1073/pnas.1511085112

9 Revolutionizing Interface Design

Leveraging Additive Manufacturing and User-Centered Assessments

Zahra Ansari, Sajjad Rostamzadeh, and Alireza Abouhossein

INTRODUCTION

Two major causes of occupational musculoskeletal disorders are exposure to repetitive submaximal exertions and maintaining awkward posture [1]. In many industrial-related jobs, individuals perform a task using a handheld tool or devices by exerting repetitive and sustainable exertions to achieve effective results. The lack of ergonomic design in such handheld tools creates long-term trauma, resulting in lasting musculoskeletal disorders [2]. Ergonomic design entails the thoughtful integration of the psychological, morphological, physiological, and biomechanical characteristics of users when utilizing handheld tools [3]. In physical ergonomics, the primary focus of numerous designs is incorporating human anthropometry and user biomechanics to create handheld devices with optimal morphological acceptance. In the past, a design should have been generalized based on a certain anthropometric and biomechanical trait and average human; therefore, it was not possible to be personalized based on individual differences of humans.

The advancement of additive manufacturing (AM), which is part of Industry 5.0, and artificial intelligence (AI), robotics, and assistive wearable robotics, which were enveloped in Industry 4.0 have stepped up personalized design into a new dimension increasing both productivity and safety and well-being of the user [4]. In the realm of Industry 5.0, personalized design has been further enriched through the integration of a human-centered approach, which includes the concept of user-centered design (UCD). This design philosophy places the user and their specific needs at the forefront of each stage of the design process. By adopting a UCD methodology, designers can create products, systems, and environments that are not only tailored to the unique requirements of individual users but also offer enhanced usability, accessibility, and overall satisfaction. This user-oriented approach emphasizes collaboration and iteration, ensuring that the end product is both functional and responsive to the

DOI: 10.1201/9781003625360-9

user's needs and preferences. Consequently, the human-centered focus of Industry 5.0 fosters innovation and efficiency while simultaneously catering to the diverse needs of the individuals who interact with these designs.

AM, or 3D printing, has significantly expanded its applications in product and prototype design since its invention. This advanced manufacturing technique enables the creation of complex geometries and customized designs that might not be achievable through traditional manufacturing methods. As a result, AM has significantly impacted industries, especially fields such as health care and medicine [5], assistive wearable technologies (AWT) [6], and physical human–machine interfaces (pHMI) [7], such as hand tool designs. AM offers numerous advantages, including reduced material waste, environmentally friendly materials, shorter production times, and lower production costs. By leveraging the capabilities of AM, designers and engineers can rapidly develop, test, and refine prototypes, accelerating the product development cycle and facilitating innovation. Furthermore, the versatility of AM allows for the production of lightweight structures, functional parts, and personalized products, enhancing overall design flexibility and efficiency.

Following the design of a hand tool, it is essential to establish a series of both subjective and objective metrics to evaluate its efficacy. Subjective metrics [8–10] are commonly employed in assessing ergonomically designed hand tools, as they focus on the user's perception and user experience of the tool's integration with their body. Conversely, objective metrics are derived from collected data, such as electromyography (EMG) [11], handgrip strength [12–14], electrocardiography (ECG) [15], electroencephalography (EEG), and other physiological signals, which provide quantitative insights into the user's interaction with the tool.

This chapter focuses on two distinct hand-related devices, both of which have been ergonomically designed with the intention of enhancing user experience and performance. The first device is engineered to prevent the spread of viruses and bacteria in public spaces and hospitals, while the second innovation is a crochet hand tool specifically crafted to help handcrafters reduce musculoskeletal disorders typically associated with the art of crocheting. All of these are implemented in a 3D printer. In the following sections, we will delve into the key features, functionalities, and benefits of these ergonomically driven designs, highlighting how they address the unique challenges faced by users in their respective contexts.

PROJECT 1: HANDS-FREE DOOR HANDLE

The COVID-19 pandemic has highlighted various challenges, one of which is the need for a hands-free solution to open and close doors, particularly in hospital wards. The risk of contracting the virus through contact with contaminated surfaces led to increased stress among medical professionals, who were constantly seeking ways to minimize hand contact with door handles. The development process and details of the hands-free door handle (HFDH) designed for typical Iranian hospitals can be found in a published article [7].

In response to this challenge, the development of a HFDH system became a priority to help reduce transmission risks and improve safety for medical crews and patients alike. A HFDH would allow hospital staff to enter and exit wards without the need for direct contact, thereby decreasing the likelihood of infection through surface contact. This innovative solution would not only enhance the overall hygiene within healthcare facilities but also alleviate some of the stress faced by frontline workers during these unprecedented times.

In environments with heightened risks of virus transmission, two solutions are usually considered: personal protective equipment (PPE), which plays a crucial role in reducing and potentially stopping the spread of viruses, as highlighted by Mantelakis et al. [16]. However, wearing full PPE gear can be uncomfortable and may not be necessary for everyone. Additionally, donning full-protective gowns with protective shields is time-consuming and often hampers users' dexterity. On the other hand, engineering control approaches, such as devices that minimize risk factors, can contribute to reducing virus transmission by limiting direct contact with contaminated surfaces [17]. In this context, addressing frequently touched surfaces like doorknobs, which are more susceptible to harboring viruses and bacteria, becomes essential. By implementing innovative solutions such as HFDH devices, the likelihood of virus transmission from one person to another can be significantly decreased, thereby enhancing overall safety and hygiene within various environments, including healthcare facilities, public spaces, and other high-traffic areas.

When we started the HFDH project, we realized that in Iran, hospital door handles often differ from the standard doorknobs recommended by the World Health Organization (WHO) and the International Health Facility Guidelines Part C (iHFG) (2015), which are commonly found in North American and European countries. Hospital door handles in Iran are typically made from a composite of synthetic materials and aluminium-based metal alloys. These handles, which require pushing or pulling to operate, can pose challenges in maintaining hygiene and preventing the spread of infections in healthcare settings.

The nonstandardized door handles found in Iran may present difficulties in maintaining optimal hygiene levels, highlighting the potential benefits of adopting internationally recognized standards for door handle designs in healthcare environments. During our visit to three major hospitals located in the North and Northeastern regions of Tehran, affiliated with Shahid Beheshti University of Medical Sciences (SBMU) – Iran's second largest and oldest teaching and research university – we observed a lack of standardization in door handle designs; see Figure 9.1. The doorknobs installed in these hospitals exhibited significant variations, indicating that they do not adhere to any specific standard.

Designing an ergonomic hand tool involves incorporating UCD principles to ensure the final product addresses the user's needs, limitations, and preferences. This approach considers the user's physical capabilities, cognitive processes, and the context in which the tool will be used.

A typical UCD process involves the following steps:

1. **Understanding the user**: Gather data on the target users, including their physical characteristics, strengths, limitations, and how they currently perform the task.
2. **Identifying user needs and pain points**: Determine the specific requirements and challenges the user faces when using existing tools.

FIGURE 9.1 Typical door handles in three hospitals affiliated with SBMU [7].

3. **Designing the hand tool**: Create a tool design that addresses the user's needs and limitations while improving comfort, safety, and efficiency.
4. **Evaluating and iterating**: Test the design with users, gather feedback, and make any necessary revisions. This iterative process helps refine the design and increase user satisfaction.

By focusing on UCD principles, ergonomic hand tools can be developed that not only meet functional requirements but also provide optimal comfort, safety, and efficiency for the individuals who use them.

In designing the HFDH, several key criteria were considered to ensure its effectiveness and usability within hospital settings:

1. **Universal compatibility**: The HFDH should be adaptable to any door handle type found in hospitals, allowing for seamless integration with existing door handle infrastructure.

2. **Inclusive design**: The HFDH should be designed to cater to the anthropometric dimensions of both men and women, ensuring ease of use and accessibility for individuals with varying hand sizes.
3. **Proprioception sensation**: The HFDH should incorporate elements that provide subtle cues to users, enabling them to locate and operate the door unconsciously, without needing to actively search for it. This enhances overall user experience and efficiency when navigating hospital environments.

By addressing these essential criteria, the HFDH can be designed as a user-friendly, efficient, and adaptable solution for hands-free door operation in hospital settings.

Our research group has initiated the research and development of proprioceptive insoles using AM, which are specifically designed to stimulate the neuromuscular proprioception of individuals who require enhanced sensory feedback at the soles of their feet. These insoles aim to provide users with improved balance, stability, and overall proprioceptive awareness, which can be particularly beneficial for individuals with neurological conditions or those recovering from injuries. The proprioceptive insoles are engineered with texture at the top causing sensory stimulation features that target specific nerve receptors in the feet. This innovative technology has the potential to significantly enhance users' motor control, gait, and overall confidence during movement, ultimately improving their quality of life and promoting greater independence in daily activities. As an example, Figure 9.2 depicts our research group has developed a proprioceptive insole specifically designed for individuals with type 2 diabetes who often experience a loss of sensation in their feet due to

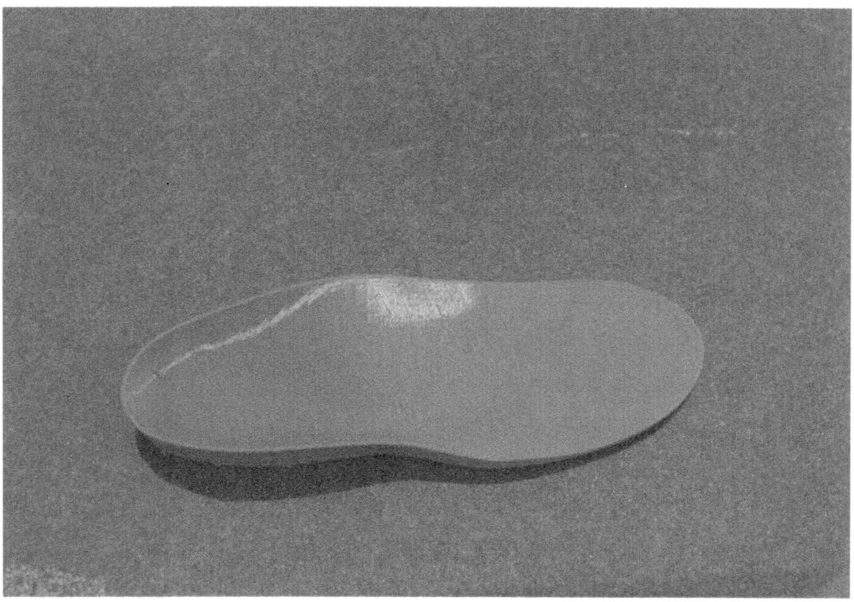

FIGURE 9.2 Proprioceptive insole made using AM in our laboratory.

diabetic neuropathy. This innovative insole aims to revive and enhance the sensory feedback that is typically diminished in these patients.

FINAL DESIGN

The next step was to establish the required criteria to design the HFDH using any computer-aided design (CAD) software, with the average using 95 percentile forearms from Iranian women [18]. Once that is done, the file can be saved in stereolithography (.stl) format. That file should be used to slice the .stl file and generate the required G-code using any of the software *Cura*, *Slic3r*, and *Simplify3D* [19, 20]. The first two are free, and the last one you can purchase with a single payment. Once the designed G-code is produced, the part can be generated in the 3D printer. The final product is shown in Figure 9.3.

Our research team successfully designed and manufactured a HFDH using 3D printing technology. The design was based on the 95th percentile of Iranian forearm anthropometric data, ensuring that it accommodates the majority of Iranian hospital personnel. This approach was particularly important since door handles in hospitals often do not adhere to any specific standard.

The HFDH was assessed for system performance using the System Usability Scale (SUS) questionnaire, comparing it to existing door handles in hospital wards. The results demonstrated that the usability of our HFDH design was significantly higher than that of the preexisting door handles (85% vs 56%). This finding highlights

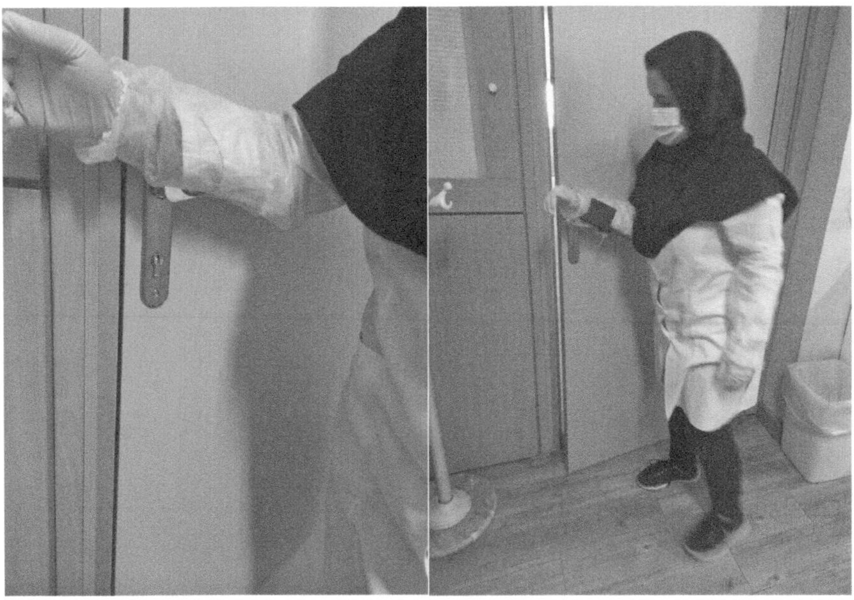

FIGURE 9.3 Ordinary door handle in the hospital complex right: HFDH produced in our laboratory by the AM process used to open hospital doors.

the potential of our ergonomic and UCD approach in creating more efficient, user-friendly, and hygienic door handles for healthcare environments.

By addressing the needs of hospital staff and considering local anthropometric data, our hands-free ergonomic door handle has the potential to enhance safety, accessibility, and overall user satisfaction in hospitals, thus contributing to improved working conditions and patient care.

PROJECT 2: DESIGN OF CROCHET HOOK HANDLE

Crochet is a type of knitting that was originally done by Scottish peasants using a small hooked needle called a shepherd's hook. The peak of this art's perfection dates back to England in the early 19th century. One of the differences between crochet and knitting is that in crochet, each stitch is completed before starting the next one, whereas in knitting, multiple open stitches are held on the needles simultaneously. In right-handed crochet, the person holds the crochet hook with their right hand and the yarn with their left hand, and for left-handed individuals, it is the opposite. The way to hold the hook in hand is generally either like a knife or a pencil, meaning with an internal or external precision grip, respectively. Additionally, depending on habit and skill level, individuals can hold the hook in various styles within these two methods. This means depending on how they hold the crochet hook handle (CHH) certain areas of the hand may feel discomfort.

Crocheting without proper support for the forearm and elbow often leads to poor posture, including a hunched neck and bent, ulnar-deviated wrists. This issue is prevalent regardless of the type of hook used during the crocheting process. As the hand and hook function as a single system, particularly with a knife grip, any modifications to the hook's specifications mainly affect the hand and fingers but have minimal impact on the posture of other parts of the upper limb [15]. To address postural concerns during crocheting, various corrective measures can be implemented. Specially designed cushions can provide targeted support for the arm and forearm, helping to alleviate strain on the neck and elbow while promoting a more upright and ergonomically sound posture. Additionally, soft exoskeletons for the shoulder may offer enhanced support and reduce the risk of musculoskeletal disorders. However, such solutions may not be economically feasible for all users and could impose functional constraints that limit user comfort and mobility. By implementing these posture-correcting interventions, individuals engaging in crocheting can experience improved comfort, reduced strain on their upper limbs, and a decrease in the potential for long-term injuries or discomfort associated with poor posture during the activity [16].

In addition, designing an ergonomically optimized CHH plays a pivotal role in mitigating the discomfort often experienced by users in their thenar and hypothenar muscles during submaximal exertion. These muscles, located at the base of the thumb and little finger, respectively, are heavily involved in the intricate movements required for crochet work.

An ergonomic handle that conforms to the natural contours of the hand and supports proper finger placement reduces strain on the thenar and hypothenar muscles. This improved hand posture decreases muscle fatigue, cramping, and other discomfort associated with extended periods of crocheting.

Consequently, the implementation of an ergonomic CHH not only enhances user comfort and overall satisfaction but also promotes long-term hand health for avid crocheters. By minimizing strain on the thenar and hypothenar muscles, such a design fosters a more enjoyable and sustainable crochet experience for users of all skill levels. This section reports on the design of a ergonomically crochet handle using 3D printer using soft polylactic acid (PLA) material. The present study consisted of two primary stages: design and assessment. The design phase employed an iterative process, integrating ergonomic principles and anthropometric hand measurements to develop a prototype of the CHH using SolidWorks software. This prototype was then transformed into stereolithography (.stl) format, enabling the production of a 3D-printed crochet handle, designed to hold a hook, utilizing soft PLA material. During the assessment phase, the redesigned shape of the CHH was assessed for its impact on subjective variables by engaging 36 participants. The findings from this comprehensive approach provided valuable insights into the effectiveness of the ergonomically enhanced CHH.

To develop a human-centered CHH, a decision matrix was employed to effectively consolidate and analyze information from various sources. This matrix helped identify and prioritize essential factors that should be considered when designing any tool including a hand tool, specifically for the crochet profession.

The decision matrix was constructed based on the following steps:

1. **Identifying potential factors**: A comprehensive list of factors that could potentially impact the design of a CHH was compiled from existing literature and expert opinions.
2. **Weighting the factors**: Each identified factor was assigned a weight based on its perceived importance in designing a CHH. The weights were determined by consulting experts in the field.
3. **Analyzing expert opinions**: The views and recommendations of professionals experienced in crochet and hand tool design were collected and analyzed, focusing specifically on the crochet profession.
4. **Populating the matrix**: The identified factors and their corresponding weights were entered into the decision matrix. Expert opinions were then used to rate each factor, creating a comprehensive overview of the most critical aspects to consider in the design process.

The use of a decision matrix in this context allowed for a systematic and objective approach to identifying and prioritizing key factors in the design of a CHH. This method enabled the research team to develop a human-centered design that addresses the specific needs and challenges faced by crochet professionals, ultimately leading to a more efficient, comfortable, and user-friendly tool (refer to Figure 9.4).

In the process of designing a CHH with a human-centered approach, nine crucial criteria were identified as the most significant factors to consider:

1. **Handle length**: The length of the handle should be optimized for user comfort and control during crochet work.

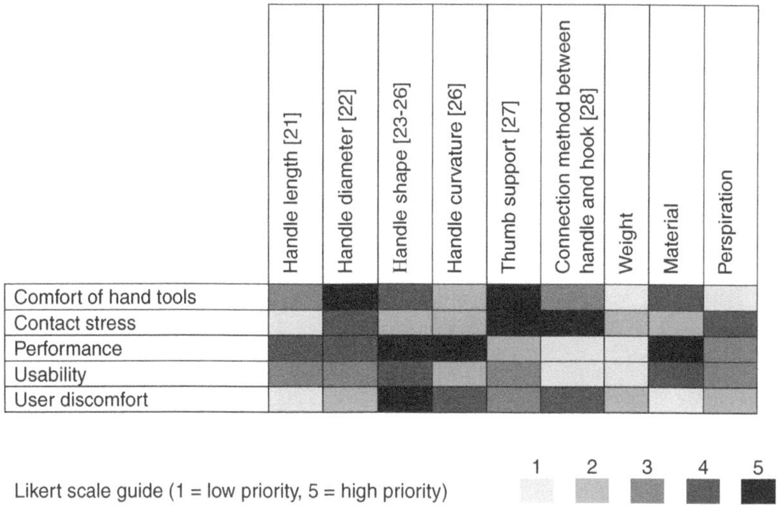

FIGURE 9.4 The initial decision matrix for starting the design process.

2. **Handle diameter**: An appropriate handle diameter ensures a comfortable grip and reduces hand strain.
3. **Handle shape**: The shape of the handle should promote a natural and ergonomic grip for improved comfort and control.
4. **Handle curvature**: The curvature of the handle should accommodate the natural position of the hand, reducing strain on the wrist and fingers.
5. **Thumb support**: The design should include a feature to support the user's thumb, providing additional stability and control.
6. **Joining strategy between handle and hook**: The method of connecting the handle to the hook should ensure a secure and durable bond while maintaining comfort and ease of use.
7. **Weight**: A lightweight design reduces user fatigue during prolonged periods of crocheting.
8. **Material**: The choice of material impacts the durability, comfort, and overall feel of the CHH.
9. **Perspiration**: The handle design should account for users' hand perspiration to maintain a secure grip and prevent slippage during use.

By considering these nine key factors in the design process, the CHH can be optimized for user comfort, control, and overall satisfaction. This holistic approach to design ensures that the final product meets the needs and preferences of crocheters, enhancing their overall experience and reducing the risk of strain or injury associated with prolonged crochet work. The results obtained from the decision matrix were used as a guide for the initial design sketch of the CHH, focusing on prioritized items at the start of the design process.

Throughout the iterative design process, CAD software such as SolidWorks [29] was used to create each model, and subsequently, it was constructed using an

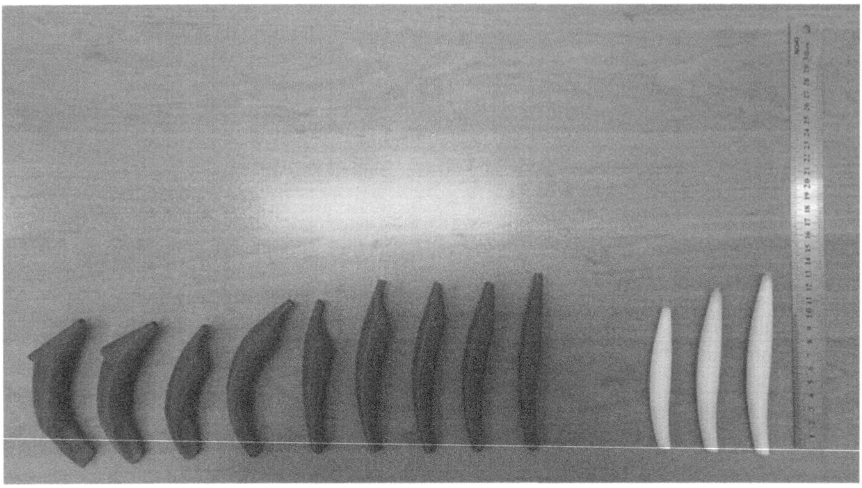

FIGURE 9.5 Printed samples of the new CHH design, arranged from the first design on the left to the final prototype on the right.

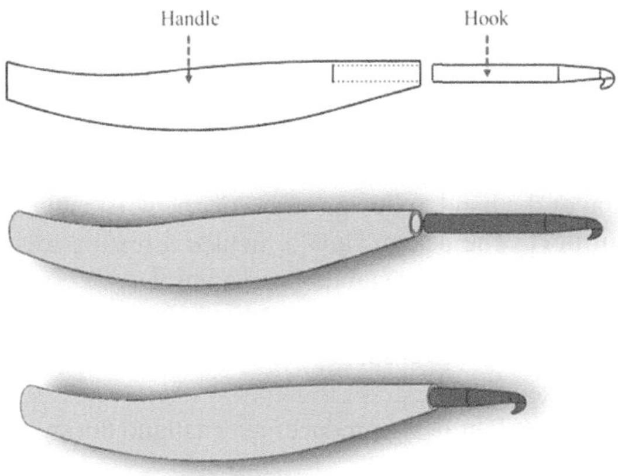

FIGURE 9.6 Assemble the hook to the handle.

in-house 3D printer. Both an expert and a novice in crochet were then invited to trial the prototype and provide feedback on its strengths and weaknesses. This valuable input, combined with established standards for internal precision grips and related tools, was instrumental in refining the design and shaping the final prototype for a crochet hook with a 4.5 mm hook size. Figure 9.5 offers a visual representation of the evolution of CHH, from the initial design to the final prototype, showcasing the various iterations and improvements made along the way.

Given that the primary objective of this study was the ergonomic design of the CHH, a ready-made hook was employed for this purpose. The handle was designed

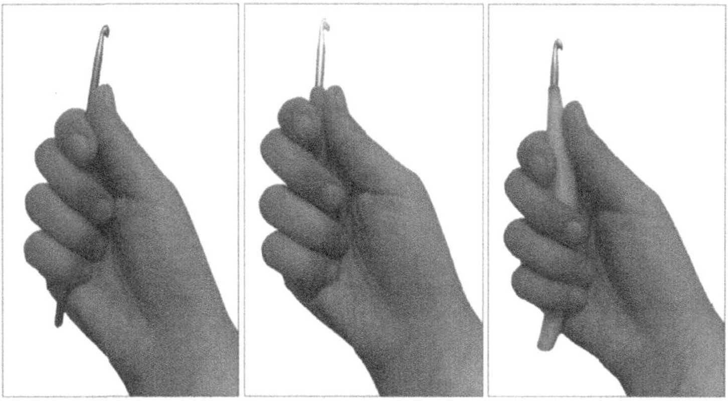

FIGURE 9.7 Comparison of the grips of the crochet hooks evaluated in the study: (a) new handle, (b) silicone handle, and (c) aluminum handle.

to accommodate the hook with a clearance fit by incorporating a hollow cylindrical space with an appropriate angle at the front part of the handle (Figure 9.6). This approach allowed the researchers to concentrate on optimizing the handle design while ensuring compatibility with standard crochet hooks.

Crocheting is a precision-oriented activity that necessitates an ideal balance of performance and comfort. A crucial factor in achieving this balance is the positioning of the thumb on the CHH. The thumb's dynamic nature during crocheting involves longitudinal movement along the handle, depending on the type of stitch being used. This movement causes the thumb to shift closer to and away from the hook intermittently. To accommodate this dynamic nature, several CHHs were designed and constructed, and after thorough experimentation, the optimal angle between the hook and the handle was determined to be 15 degrees. Figure 9.7 depicts the grip of the crochet hook and the finger placements on the handle in the final prototype, along with the two conventional crochet hooks evaluated in the study, providing a visual comparison of the ergonomic enhancements made to the design.

The final prototype's CHH features a consistently circular cross-section along its entire length, eliminating sharp edges to minimize stress concentration on the palm and fingers. This design also aids in reducing the torque required for task execution [30]. To further ensure even force distribution across the palm during crocheting, the handle was crafted with a tapered shape. This ergonomic design facilitates a comfortable and secure grip by the fingers while reducing the need for excessive gripping force. The handle's cross-section has a smaller diameter where the little finger grips it, taking into account its shorter length and reduced ability to wrap around the handle. Conversely, the cross-section boasts the largest diameter, where the middle finger, being the longest, holds the hook, providing a customized and comfortable grip for each finger.

TABLE 9.1
Types of handles evaluated in the study

Code	Handle type
A	Aluminum
B	Silicone
C	New handle

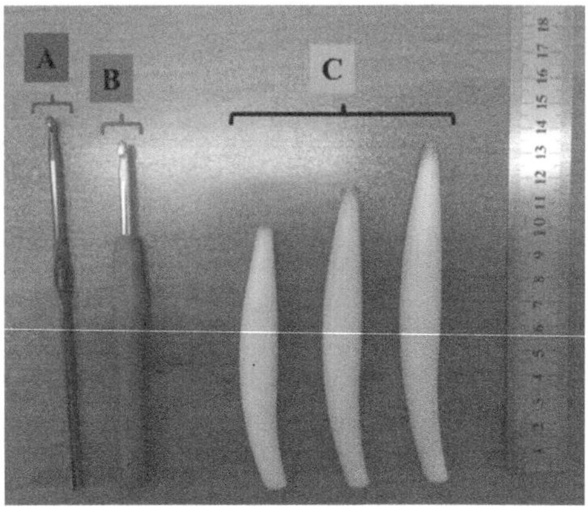

To accommodate a range of hand sizes, the final prototype was made available in three distinct radii and lengths. These dimensions were determined by considering the anthropometric data of metacarpal hand breadth at the 5th, 50th, and 95th percentiles of Iranian women, ensuring inclusivity and ergonomic comfort for various users. Table 9.1 provides an overview of the crochet handles types evaluated throughout the study.

The assessment phase employed several assessment tools, including the comfort questionnaire for hand tools (CQH), SUS, and discomfort-based hand map. Moreover, crocheting performance was evaluated by measuring the total number of chains woven within 10 minutes. Each participant was tasked with completing crochet work using all three types of handles: the newly designed handle, silicone handle, and aluminum handle. Following each session, the participants were given questionnaires to assess the performance of the designed grip in comparison to the existing commercial grips used in crocheting. Table 9.2 provides an overview of the specifications for all crochet handles designed with the different forearm percentile of Iranian women [18].

The findings of the study revealed that the morphological modifications incorporated into the design of the new hook handle contributed to a notable reduction in discomfort and contact stress across various hand regions when compared to traditional crochet hooks. As per the CQH questionnaire, the comfort scores were 59.53, 18.92, and 97.14, while the SUS questionnaire yielded average usability scores of 59.93, 71.74, and 85.42 for the aluminum handle, silicone handle, and new handle, respectively. In both instances, the new hook handle demonstrated a significant advantage over the conventional hooks ($p < 0.05$). Conversely, the average crochet performance was measured by the number of chains completed within 10 minutes

TABLE 9.2
Characteristics of the hooks used in the study

Hook type	Hand breadth (mm)	Handle length (mm)	Crochet length (mm)	Weight (g)	Handle diameter (mm)	The angle between the hook and the handle (degree)	Handle material
Aluminum	-	-	151	6	4.5	0	Aluminum
Silicone	-	103	138	13	14–10	0	Silicone
New hook (small size)	63 (5th)	104	128	12	6.5–16	15	Soft PLA
New hook (medium size)	75 (50th)	119	143	14			
New hook (large size)	86 (95th)	132	156	16			

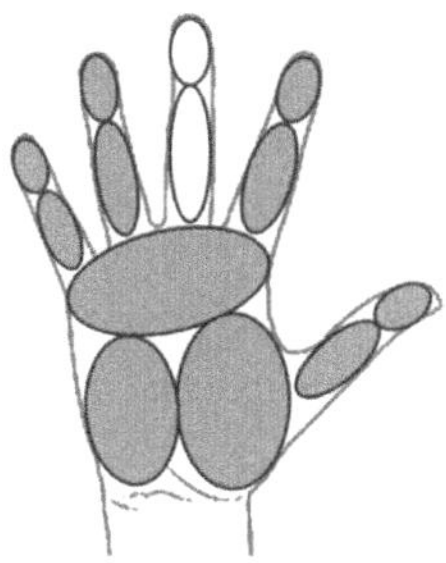

FIGURE 9.8 Users express less discomfort and contact stress when using Hook C compared to Hook A in the green areas.

which were 191, 207, and 198 chains using the aluminum handle, silicone handle, and a newly designed handle, respectively. The findings revealed that the new handle slightly underperformed in comparison to the silicone silicone handle; however, this does not indicate any significant differences between the average performance of the silicone and the newly designed hook handle ($p < 0.05$). Furthermore, the hand map results demonstrated that the new hook handle significantly alleviated contact stress in most hand and finger areas compared to both conventional hooks ($p < 0.05$) (Figures 9.7–9.8). This may be achieved because of the designed hook morphology that is curved and tapered. Among the conventional handles, the silicone handle was found to provide greater comfort than the aluminum handle ($p < 0.05$) (Figure 9.9). This may be explained because of the grip size of the silicone hook in comparison to the conventional hook, which provides a larger grip size (Figure 9.10).

The results of this study indicated that if the hand tools are designed to match the specific task with proper anthropometric dimensions they can reduce discomfort and contact stress but also enhance the usability and comfort of the user. However, further studies are needed to assess their impact on performance.

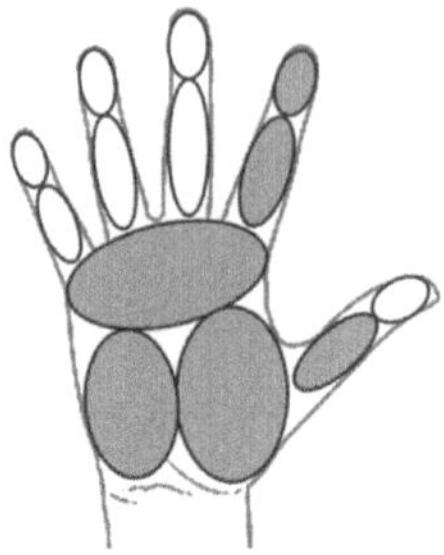

FIGURE 9.9 Users express less discomfort and contact stress when using hook handle C compared to hook handle B in the green areas.

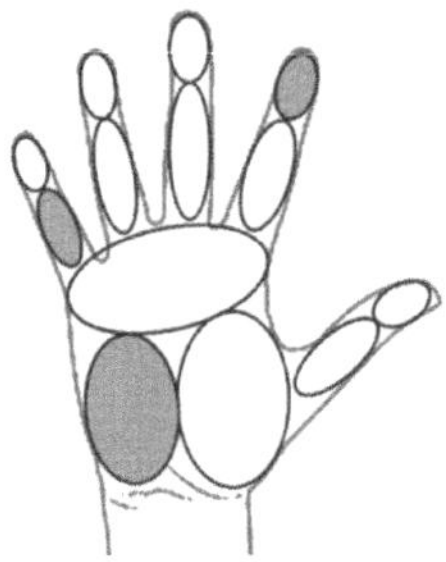

FIGURE 9.10 Users express less discomfort and contact stress when using hook handle B compared to hook handle A in the green areas.

ADJUSTING 3D PRINTER PARAMETERS TO ACHIEVE OPTIMAL OUTCOME

Each slicer software has numerous parameters, and alternating them will affect the quality of the constructed 3D parts significantly. To maintain stability during the printing process and prevent parts from falling or wobbling, it may be necessary to add supports to areas that are prone to collapse. This requirement may vary depending on the printer, as advanced models may not need additional support structures. Although any 3D printer slicer software can automatically choose to support based on printing conditions, it is essential not to rely solely on these automated selections. Successful placement of supports for printing parts is significantly influenced by practical experience in 3D printing. With a skilled operator, optimal support placement can be achieved, ensuring the integrity and quality of the printed components.

The design of the component should minimize the need for support during printing. Although the supports are removed from the main piece after printing, the connecting points where the supports are attached shall become rough and require smoothing. This process affects the surface texture and visual aesthetics of the component. The time required to print any component is also crucial. If the printing process is prolonged, it may be interrupted due to various reasons such as power outage, depletion of filament requiring a new roll to be loaded onto the printer, or even filament breakage during the process. Even if a 3D printing process can be resumed after an interruption, the cooling and reheating of the nozzle during subsequent layers can negatively impact the overall print quality. This is because the temperature variations may lead to inconsistencies in the deposition of material, potentially resulting in structural defects, layer delamination, or poor surface finish.

However, once the printing process resumes and the nozzle returns to its normal operating temperature, the print quality typically improves and approaches the quality observed before the interruption. This recovery is due to the stabilizing effect of maintaining a consistent nozzle temperature throughout the rest of the printing process.

To minimize the impact of interruptions on print quality, it is crucial to address the root cause of the interruption and strive to maintain a stable printing environment. Implementing measures such as ensuring a reliable power supply, monitoring filament feed, and maintaining consistent ambient conditions can help reduce the likelihood of interruptions and promote optimal print quality. Each material that can be used in a 3D printer has specific temperature conditions (preheating and printing temperature) and speed requirements, all of which impact the quality of the final component. Selecting the appropriate temperature range and printing speed for each type of filament not only relies on the information provided by the 3D printer manufacturer and the details on the filament roll but also directly correlates with the researcher's experience. Other important factors in 3D printing include the infill percentage of the component and its orientation placement in the print area. The infill percentage, which determines the density of the material inside the component, directly impacts its strength, weight, and print time. Proper orientation of the component on the print bed is crucial for ensuring optimal print quality and minimizing the number of supports required to keep the part upright during the printing process.

In summary, successful 3D printing depends on a combination of proper component design, material selection, and process management. Considering these factors and gaining experience in 3D printing techniques will help ensure the best possible outcomes in terms of print quality and final component properties.

DISCUSSION

One of the innovations of the last study was considering three different sizes of CHHs for the Iranian women population. Based on the results obtained, the initial sizing of the CHH was suitable for prototype design. Using a crochet hook size that fits one's hand significantly reduces contact stress in most areas of the hand, increases usability, and improves comfort. In this study, the choice of a new handle size for assessment depended on the participants' selection. It is recommended that future studies consider a range of hand dimensions that might influence handle sizing and design of patterns to provide proprioceptive feedback to the user's hand to reduce further submaximal exertion existing in repeatable work such as crochet. It is worth noting that past studies have indicated that human digits do not follow the anthropometric dimension of the hand [31]. Prior to the evaluation, the assumption was that hand width would be the primary determining factor in selecting the appropriate CHH size for participants. It was anticipated that individuals with wider hands would naturally gravitate toward larger hook handles, while those with narrower hands would opt for smaller handles. However, the study's findings did not reveal a significant positive correlation between hand width and selected hook handle length. This unexpected result can be attributed to the variation in finger length among individuals, which does not necessarily correspond to overall hand size. For example, an individual with smaller hands may have longer fingers, while someone with larger hands may have shorter fingers.

In this study, it was observed that participants tended to select CHHs based on personal preference and comfort rather than strictly adhering to designs specifically tailored to their hand size percentile. This finding emphasizes the importance of considering various factors, including individual digit length and breadth, when designing hand tools such as crochet hooks.

The results of the study suggest that hand length and breadth alone may not be sufficient for determining the most suitable tool size for an individual user. Instead, designers should also consider the unique dimensions and characteristics of each user's fingers, as these factors play a crucial role in ensuring a comfortable and ergonomic grip.

This observation highlights the need for a more nuanced approach to hand tool design, taking into account the diverse needs and preferences of users. By incorporating a wider range of anthropometric measurements and focusing on individual user comfort, designers can create hand tools that are more accessible, enjoyable, and efficient for a broader audience using AM.

These observations underscore the importance of considering not only hand width but also other hand dimensions, particularly finger length, when designing and sizing manual tool handles. Finger length measurement show poor conformity with established anthropometric relationship involving hand width and length [18]. Future studies should take into account the unique characteristics and variations of individual fingers to ensure that ergonomic hand tools accommodate a diverse range of users and provide optimal comfort, control, and efficiency. By incorporating a more comprehensive understanding of hand anthropometry, researchers and designers can contribute to the development of more inclusive and user-friendly tools.

The choice of material for the CHHs in this study was primarily determined by the limitations of the 3D printer being used. From the few materials available, such as acrylonitrile butadiene styrene (ABS) and PLA, the most suitable option was selected. As a result, the final version of the CHH was made from soft PLA filament, which provided increased flexibility and comfort compared to standard PLA. While the soft PLA material showed overall improvements in user satisfaction compared to traditional CHHs, the study did not conclusively determine the most suitable material for CHHs. This is because material selection appears to be subjective, with varying preferences among participants. Participant feedback at the end of the evaluation phase revealed a lack of consensus between silicone and soft PLA materials. This indicates that further research is needed to explore the impact of material selection on user comfort and satisfaction. Moreover, even though the silicone handle caused some hand sweating during use, its softness compared to the plastic handle was not enough to sway all participants in favor of soft PLA.

In summary, the selection of material for CHHs warrants further research, given the subjective nature of user preferences and the impact of material properties on the overall crocheting experience. By considering a wider array of materials and integrating user feedback, particularly in terms of proprioception and tactile perception, researchers can contribute to the development of CHHs that not only accommodate the diverse needs of users but also enhance their comfort, performance, and satisfaction.

A comprehensive approach to material selection should involve the following considerations:

1. **Ergonomics and Comfort:** Assess materials for their ability to provide a secure and comfortable grip, minimize strain on the hands, and accommodate different crocheting techniques.
2. **Durability and Longevity:** Evaluate materials for their resistance to wear and tear, as well as their ability to maintain shape and integrity over time.
3. **Proprioceptive Feedback:** Consider the material's capacity to provide tactile feedback that enhances control and precision during crocheting.
4. **Sustainability and Environmental Impact:** Examine the ecological implications of different materials, prioritizing options that minimize waste and pollution throughout their lifecycle. For example, PLA is biodegradable and more environmentally friendly than ABS.

By accounting for these factors and prioritizing user feedback in the design process, researchers can facilitate the development of CHHs that optimize user experience, foster inclusivity, and promote sustainability in the world of crocheting.

Given that both studies relied on subjective evaluation tools, it is highly advisable to incorporate objective measures in future research. By including data from EMG or electroencephalogram (EEG), a more comprehensive understanding of the physiological responses and cognitive processes related to the use of HFDH and CHHs can be obtained.

EMG can provide valuable insights into muscle activity and fatigue during crocheting, helping researchers assess the ergonomic efficacy of different handle designs. This information could contribute to refining the design of CHHs to reduce muscle strain and improve overall comfort for users. Additionally, EEG can be employed to examine the cognitive processes associated with crocheting, such as attention, concentration, and task-related stress. By evaluating the neurological responses related to various CHH designs, researchers can further enhance the overall user experience, ensuring that the tools are not only comfortable but also promote optimal cognitive engagement. By combining subjective assessments with objective measures like EMG and EEG, future studies can provide a more holistic understanding of the factors contributing to the comfort, usability, and overall satisfaction associated with CHH design. This multidimensional approach to research will facilitate the development of more effective and user-friendly tools that cater to the diverse needs and preferences of individuals engaged in crocheting.

ACKNOWLEDGMENTS

These works were supported by the Shahid Beheshti University of Medical Sciences, Tehran, Iran, under Grant No. 99-24474 and Grant No. 02-43005485. The study protocol was reviewed and approved by the Research and Ethics Committee of Shahid Beheshti University of Medical Sciences for these two aforementioned projects IR.SBMU.PHNS.REC.1399.131 and IR.SBMU.PHNS.REC.1402.001, respectively. The authors would like to express special thanks to all the participants who contributed to the research.

REFERENCES

[1] M. S. Hossain *et al.*, "Ergonomic risk prediction for awkward postures from 3D keypoints using deep learning," *IEEE Access*. vol. 11, pp. 812–824, 2023.

[2] W. G. Lewis and C. J. A. E. Narayan, "Design and sizing of ergonomic handles for hand tools," vol. 24, no. 5, pp. 351–356, 1993.

[3] G. Vladić *et al.*, "Investigation of factors influencing ergonomic characteristics of water bottle handles," vol. 29, no. 5, pp. 1545–1551, 2022.

[4] J. Alves, T. M. Lima, and P. D. J. P. Gaspar, "Is industry 5.0 a human-centred approach? A systematic review," vol. 11, no. 1, p. 193, 2023.

[5] S. Hellman, P. Frisch, A. Platzman, and P. J. D. H. Booth, "3D printing in a hospital: Centralized clinical implementation and applications for comprehensive care," vol. 9, p. 20552076231221899, 2023.

[6] A. Abouhossein, U. Martinez-Hernandez, M. I. Awad, I. Mahmood, D. Yilmaz, and A. A. Dehghani-Sanij, "Assistive Gait Wearable Robots—From the Laboratory to the Real Environment," in *Reinventing Mechatronics: Developing Future Directions for Mechatronics*, X.-T. Yan, D. Bradley, D. Russell, and P. Moore, Eds. Cham: Springer International Publishing, 2020, pp. 75–92.

[7] A. Abouhossein, Z. Ansari, and S. Kalkhoran, "3D printed hands-free door handle to prevent COVID-19 virus spread: Developing a design based on Iranian population for medical centers," in *AHFE International Conference*, vol. 146, 2024, pp. 171–183.

[8] K. Byungjoon BJ, "Effect of ergonomic design changes in hand tools on physiological cost and subjective ratings," *International Journal of Occupational Safety and Ergonomics,* vol. 18, no. 2, pp. 267–277, 2012.

[9] S. Graziosi, F. Ferrise, M. Bordegoni, and O. Ozbey, "A method for capturing and translating qualitative user experience into design specifications: the haptic feedback of appliance interfaces," in *DS 75-7: Proceedings of the 19th International Conference on Engineering Design (ICED13), Design for Harmonies, Vol. 7: Human Behaviour in Design, Seoul, Korea, 19–22.08. 2013*, pp. 427–436, 2013.

[10] S. Rostamzadeh, A. Abouhossein, M. Saremi, F. Taheri, M. Ebrahimian, and S. Vosoughi, "A comparative investigation of machine learning algorithms for predicting safety signs comprehension based on socio-demographic factors and cognitive sign features," *Scientific Reports*, vol. 13, no. 1, pp. 1–20, 2023.

[11] J. C. Gillette, S. Saadat, and T. Butler, "Electromyography-based fatigue assessment of an upper body exoskeleton during automotive assembly," *Wearable Technologies,* vol. 3, p. e23, 2022, Art. no. e23.

[12] S. Rostamzadeh, M. Saremi, and T. J. W. Fereshteh, "Maximum handgrip strength as a function of type of work and hand-forearm dimensions," vol. 65, no. 3, pp. 679–687, 2020.

[13] M. Saremi and S. Rostamzadeh, "Hand dimensions and grip strength: a comparison of manual and non-manual workers," in *Proceedings of the 20th Congress of the International Ergonomics Association (IEA 2018) Volume IX: Aging, Gender and Work, Anthropometry, Ergonomics for Children and Educational Environments 20*, pp. 520–529, Springer, 2019.

[14] M. Saremi, S. Rostamzadeh, M. J. and Nasr Esfahani, "Hand functionality in dentists: the effect of anthropometric dimensions and specialty," International Journal of Occupational Safety and Ergonomics, vol. 28, no. 3, pp. 1473–1481, 2022.

[15] S. Rostamzadeh, A. Abouhossein, S. Vosoughi, S. B. Gendeshmin, R. J. and Yarahmadi, "Stress influence on real-world driving identified by monitoring heart rate variability and morphologic variability of electrocardiogram signals: the case of intercity roads," International Journal of Occupational Safety and Ergonomics, vol. 30, no. 1, pp. 252–263, 2024.

[16] A. Mantelakis, H. V. M. Spiers, C. W. Lee, A. Chambers, and A. Joshi, "Availability of personal protective equipment in NHS hospitals during COVID-19: A national survey," *Annals of Work Exposures and Health,* vol. 65, no. 1, pp. 136–140, 2021.

[17] K. E. Zelik, C. A. Nurse, M. C. Schall Jr, R. F. Sesek, M. C. Marino, and S. J. A. e. Gallagher, "An ergonomic assessment tool for evaluating the effect of back exoskeletons on injury risk," *Applied Ergonomics,* vol. 99, p. 103619, 2022.

[18] S. Rostamzadeh, A. Abouhossein, K. Alam, S. Vosoughi, and S. S. Sattari, "Exploratory analysis using machine learning algorithms to predict pinch strength by anthropometric and socio-demographic features," *International Journal of Occupational Safety and Ergonomics,* vol. 30, no. 2, pp. 518–531, April 2, 2024.

[19] M. Šljivic, A. Pavlovic, M. Kraišnik, and J. Ilić, "Comparing the accuracy of 3D slicer software in printed end-use parts," in *IOP Conference Series: Materials Science and Engineering*, vol. 659, no. 1, p. 012082, IOP Publishing, 2019.

[20] A. Locker. (2019). *Best 3D Slicer Software for 3D Printers in 2019*. Available: https://all3dp.com/1/best-3d-slicer-software-3d-printer/

[21] H. Strasser, *Assessment of the ergonomic quality of hand-held tools and computer input devices*. IOS press, 2007.

[22] Y. Aldien, D. Welcome, S. Rakheja, R. Dong, and P.-E. Boileau, "Contact pressure distribution at hand–handle interface: role of hand forces and handle size," *International Journal of Industrial Ergonomics,* vol. 35, no. 3, pp. 267–286, 2005.

[23] I. Dianat, M. Nedaei, and M. A. M. Nezami, "The effects of tool handle shape on hand performance, usability and discomfort using masons' trowels," *International Journal of Industrial Ergonomics,* vol. 45, pp. 13–20, 2015.

[24] M. Motamedzade, A. Choobineh, M. A. Mououdi, and S. Arghami, "Ergonomic design of carpet weaving hand tools," *International Journal of Industrial Ergonomics,* vol. 37, no. 7, pp. 581–587, 2007.

[25] D. S. Bisht and M. R. Khan, "A novel anatomical woodworking chisel handle," *Applied Ergonomics,* vol. 76, pp. 38–47, 2019.

[26] G. Harih and B. Dolšak, "Recommendations for tool-handle material choice based on finite element analysis," *Applied Ergonomics,* vol. 45, no. 3, pp. 577–585, 2014.

[27] A. Muralidhar and R. Bishu, "Safety performance of gloves using the pressure tolerance of the hand," *Ergonomics,* vol. 43, no. 5, pp. 561–572, 2000.

[28] I. Dianat, S. Asadollahi, and M. Nedaei, "Evaluation of design alternatives for sewing scissors with respect to hand performance, discomfort and usability," *International Journal of Occupational Safety and Ergonomics,* vol. 25, no. 3, pp. 386–393, 2017.

[29] "Solidworks," ed: SolidWorks (2016). SolidWorks 2016 [Computer software]. Dassault Systèmes SolidWorks Corporation., 2016.

[30] J. O. Crawford, E. Wanibe, and L. Nayak, "The interaction between lid diameter, height and shape on wrist torque exertion in younger and older adults," *Ergonomics,* vol. 45, no. 13, pp. 922–933, 2002.

[31] M. Peters, W. Potthast, S. Wischniewski, and I. Komnik, "Wearable sensors for classification of load-handling tasks with machine learning algorithms in occupational safety and health: a systematic literature review," *Ergonomics*, pp. 1–21, 2025. https://doi.org/10.1080/00140139.2025.2486193

10 Unlocking Sustainable Urbanization

Investigating the Potential of Modular Construction and 3D Construction Printing in South Africa's Affordable Housing Sector

Jeffrey Mahachi

INTRODUCTION

The construction industry worldwide is at a critical juncture, with calls for innovation growing louder amid rising concerns about sustainability, affordability, and housing shortages. Despite an evolving regulatory environment that encourages the use of non-traditional building materials and methods, the pace of adoption for modern methods of construction (MMCs) remains slow, particularly in South Africa. MMCs, as defined in this study, include advanced construction techniques such as prefabricated products and 3D construction printing that are certified by Agrément South Africa (Agrément SA, 2010) in cases where no South African National Standards exist to evaluate their performance. These methods can potentially revolutionize the industry by addressing the country's housing backlog, delivering higher-quality construction, and offering cost-effective solutions over a building's lifecycle.

Internationally, innovative construction practices have reshaped how homes are built, enhancing material quality and building process efficiency, performance standards, affordability, and occupant satisfaction. In South Africa, the housing sector is divided into two distinct markets: the bondable housing market, driven by the private sector, and the government-subsidized housing market, aimed to provide shelter for lower-income households (Mahachi, 2021). Families earning less than R3,500 (approximately US$200) qualify for government housing subsidies designed to assist those struggling to secure adequate shelter. While the bondable housing market has experienced modest yet steady growth, the situation is starkly different in the government-subsidized sector, where deliveries have stagnated.

As illustrated in Figure 10.1, the annual number of units constructed has dramatically declined from around 240,000 in 1998 to under 50,000 by 2022. This decline

DOI: 10.1201/9781003625360-10

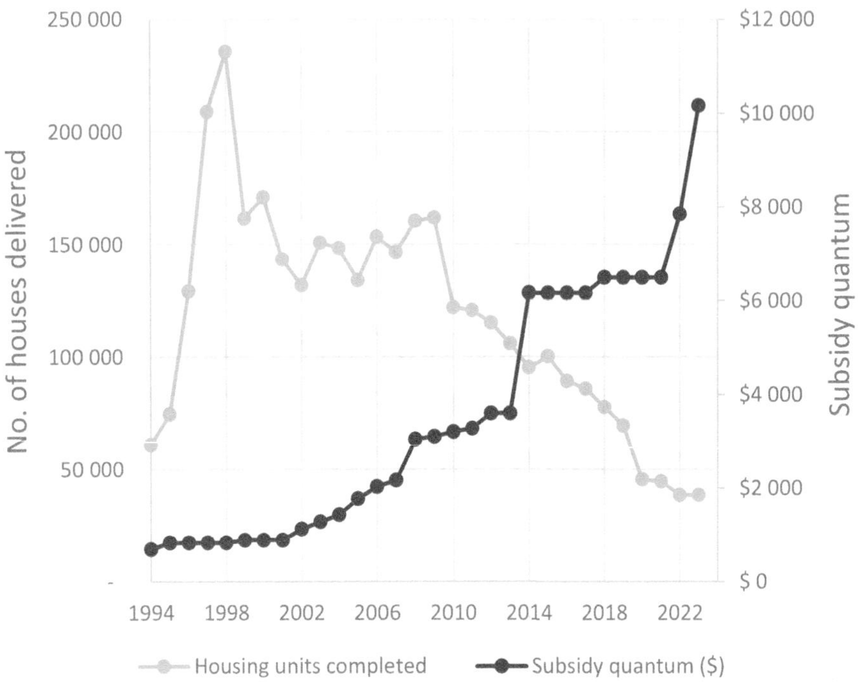

FIGURE 10.1 Delivery of houses. (Aggregated from www.dhs.gov.za).

is particularly concerning given the government's increasing investment in housing subsidies, which have risen over the years to address the growing demand for affordable housing. The figure highlights a troubling disconnect between financial support and actual construction, underscoring the urgent need for significant interventions to reverse this trend and ensure vulnerable populations can access adequate and sustainable housing.

This decline in subsidized housing delivery points to an urgent need for transformative interventions that leverage existing technical capabilities to impact social infrastructure and housing significantly. Addressing this challenge requires a paradigm shift in how housing is conceived, designed, and delivered. Integrating MMC presents a powerful opportunity to rethink traditional approaches and harness emerging technologies to meet the growing demand for affordable housing. However, these methods face several obstacles, including limited awareness, perceived high costs, and societal scepticism regarding new construction technologies.

Agrément SA (2017) plays a critical role in bridging these gaps by providing rigorous testing and certification for innovative building products, thereby ensuring that they meet the necessary performance and safety standards. Yet, the pathway from certification to widespread market acceptance remains fraught with challenges. To foster a more robust adoption of MMCs, there is a pressing need for a well-organized and comprehensive knowledge base that captures the certified performance, limitations, cost implications, and environmental benefits of these systems. Such a repository would not only facilitate better decision-making but also support the construction sector's alignment with broader sustainability objectives, including carbon emission reductions and the pursuit of energy-efficient building practices.

In this context, the study presented in this chapter aims to explore the transformative potential of MMCs and 3D construction printing within South Africa's affordable housing sector. It seeks to uncover how these methods can drive sustainable urbanization, reduce the environmental footprint of construction, and enhance the quality of life for residents. Examining the challenges and opportunities associated with innovative building systems sets the stage for a discussion on overcoming barriers to adoption and maximizing the impact of modern construction technologies (Burger, 2017). This chapter aims not merely to contribute to the body of academic literature but to spark a broader conversation on how the construction industry can evolve to meet the complex demands of a rapidly urbanizing society while supporting the nation's sustainable development goals.

AIM OF THE STUDY

The aim of this study is to explore the potential of modular construction and 3D construction printing to advance sustainable urbanization within South Africa's affordable housing sector. The research involves a detailed examination of walling technologies, with a particular focus on those assessed and certified by Agrément South Africa. It investigates variations in performance, classifies technologies based on their characteristics, identifies barriers to their adoption, and evaluates the influence of government policies on the integration of these innovations. Utilizing Agrément SA certification data and supplementary insights from technology developers, the study aims to provide a comprehensive analysis of certified walling technologies. This analysis seeks to uncover both the benefits and challenges associated with their practical application in the context of affordable housing in South Africa.

RESEARCH METHODOLOGY AND DESIGN

This study employs a structured and systematic four-phase research methodology to ensure scientific rigor and reliability. The initial phase involves an in-depth review of building performance standards and the certification process used by Agrément SA for walling systems Agrément (2010). This step provides essential insights into the evaluation criteria and processes.

The second phase is exploratory, focusing on gathering data from Agrément SA's publicly available resources and engaging with officials to analyse certification documents. This data is used to categorize walling systems based on their performance characteristics.

The third phase validates the findings by engaging directly with technology developers and system owners. This process verifies the accuracy of certification details while capturing additional perspectives on their experiences with implementing innovative technologies.

In the final phase, the study synthesizes the collected information to develop a conceptual framework that supports further academic and practical analysis. This multiphase methodology ensures that the study generates reliable, well-validated insights, contributing meaningfully to understanding walling technologies in South Africa's affordable housing sector.

LITERATURE REVIEW

The adoption of MMC has gained traction globally, with countries facing distinct challenges and opportunities in integrating these innovative practices. This literature review examines the hurdles and advancements in MMC implementation across various regions, including developed and developing nations.

In the United Kingdom, the construction industry has taken significant steps towards MMC integration, driven by government initiatives and increasing recognition of benefits such as reduced project timelines and improved sustainability. However, challenges persist, including resistance from traditional construction stakeholders, high upfront costs, and a fragmented regulatory landscape (Pan et al., 2018; Lützkendorf & Balouktsi, 2017).

Across Europe, the approach to MMC varies widely. Nations like Germany and Sweden have adopted off-site construction methods more readily, while others face obstacles such as inconsistent standards and interoperability issues (Ling et al., 2020). Initiatives like Horizon 2020 aim to address these barriers by fostering collaboration and innovation within the sector.

In the United States, MMC adoption is hindered by a predominantly site-based construction culture, limited standardization, and complex regulations (Ogunlana et al., 2018). Despite these challenges, breakthroughs in 3D construction printing are gaining momentum, with projects like ICON's affordable housing developments demonstrating the technology's transformative potential (Buswell et al., 2018).

Japan, renowned for its technological advancements, faces unique challenges in adopting MMC. An ageing workforce and a strong focus on traditional craftsmanship present barriers to widespread implementation (Kaneko et al., 2018). Nonetheless, government-led efforts to boost research and development are positioning Japan as a leader in innovative construction practices (Wong et al., 2019).

In Australia, MMC adoption is slowed by skill shortages and a lack of standardized procedures (Love et al., 2019). However, the government's focus on sustainability, and efficiency has spurred initiatives that promote modular construction (Ma et al., 2018).

China's rapid urbanization has created an urgent need for MMC to address housing demands. Key challenges include quality assurance, industry fragmentation, and insufficient investment in research and development (Li et al., 2020). The government's strong backing, particularly for 3D construction printing, reflects a commitment to modernizing the sector (Yuan et al., 2018).

Developing nations face distinct hurdles in MMC adoption, shaped by resource constraints and varying levels of technological readiness. In Malaysia, for example, low awareness and adherence to traditional construction methods limit the uptake of MMC practices (Haron et al., 2020).

3D construction printing, as an emerging technology, represents a paradigm shift in the construction industry. Despite its potential for cost-effectiveness and sustainability, challenges such as material limitations, regulatory ambiguities, and the lack of standardization must be addressed (Gibb & Shanks, 2019). Advances in robotics and innovative materials offer promising pathways to overcome these barriers (Khoshnevis et al., 2016).

Building performance as a concept has undergone substantial evolution, with notable contributions from scholars such as Foliente (2000), Fairclough (2002), and Hartkopf et al. (2008). Performance-based standards, which focus on outcomes rather than prescriptive methodologies, align with the definition of performance-based buildings as outlined by the CIB (2003). This approach has influenced modern construction practices, providing a framework for assessing building technologies and their contribution to sustainability goals.

The South African building regulatory framework embraces a performance-oriented approach that aligns with international best practices for ensuring construction safety, durability, and efficiency Agrément (2010). Central to this framework are the National Building Regulations (NBR) and Building Standards Act (Act No. 103 of 1977, as amended), which outline three distinct avenues through which construction projects can demonstrate compliance with performance requirements. These avenues provide flexibility in the approval process, enabling different methods of construction to be evaluated based on their ability to meet the functional standards set by the regulations.

The first avenue is the deemed-to-satisfy route, which allows for compliance by following prescribed building solutions specified in the South African National Standards (SANS). Under this approach, building materials and methods are considered to meet regulatory requirements automatically if they conform to established standards. This route is most suitable for conventional construction practices where standardized methods and materials are used, ensuring predictability and consistency in meeting the minimum performance criteria. However, it may not accommodate innovative or nontraditional building solutions that fall outside the established standards, limiting its applicability to new construction technologies.

The second option, known as rational design and assessment, provides an alternative pathway where compliance is demonstrated through a professionally engineered design. This approach is typically employed when building elements or methods do not fall entirely within the scope of deemed-to-satisfy standards. In such cases, a competent person, such as a registered professional engineer, prepares a rational design that demonstrates how the construction will satisfy the functional requirements of the regulations. This method allows for greater flexibility and accommodates more complex or innovative solutions if the design is backed by sound engineering principles and calculations that ensure the safety and functionality of the structure.

The third avenue is the performance-based route, which requires independent assessment and certification by Agrément SA. This approach is primarily used for MMC and innovative building products that do not conform to conventional standards or established design methods. Through a rigorous evaluation process, Agrément SA assesses the performance characteristics of the construction system, including aspects such as structural strength, fire resistance, and thermal performance. Products and systems that meet the required standards are granted Agrément certification, indicating that they have been tested and approved as fit for purpose. This certification is essential for technologies like 3D construction printing and modular construction, which often introduce novel materials and methods not covered by traditional building codes.

By providing these three distinct avenues—deemed-to-satisfy, rational design, and performance-based assessment—the South African building regulatory framework accommodates a broad range of construction techniques and encourages innovation while maintaining safety and quality standards. Each pathway offers a unique approach to compliance, allowing builders, designers, and innovators to choose the most suitable method for their specific project requirements.

In summary, the obstacles to implementing MMC and 3D construction printing are complex and differ widely between countries, influenced by unique socio-economic conditions, cultural factors, and regulatory frameworks. Gaining a thorough understanding of these challenges is crucial for devising targeted strategies that can facilitate the widespread adoption of innovative construction technologies worldwide.

REVIEW OF BUILDING PERFORMANCE AND AGRÈMENT SA CERTIFICATION

This section provides a detailed evaluation of various walling systems certified by Agrément SA, focusing on those currently active in the database. A strategic sampling methodology, guided by Agrément SA, is employed to select and analyse the systems, which are categorized based on structural performance into six main groups: sandwich insulated panels (SIPs), light concrete building systems, polyvinyl chloride (PVC) building systems, container building systems, building blocks, and 3D construction printing. Each system exhibits unique characteristics that cater to different construction needs, offering innovative solutions for enhancing building performance, sustainability, and efficiency.

1. **Sandwich Insulated Panels (SIPs)**

   SIPs consist of multiple layers that are bonded together to form a composite building material. Typically, these layers include an insulating core made of foam (such as expanded polystyrene or polyurethane) sandwiched between two rigid outer layers, often made of oriented strand board (OSB), fibre cement, or metal sheets. This layered structure results in a strong, lightweight material that provides excellent thermal insulation, making it ideal for energy-efficient buildings. Because of their inherent strength, SIPs can also contribute to the structural integrity of a building, allowing for reduced reliance on traditional framing materials.

   SIPs are especially suitable for off-site manufacturing and modular construction. The panels can be prefabricated in a controlled factory environment, ensuring high quality and consistent performance. Once completed, the panels are transported to the construction site for rapid assembly, significantly reducing construction time compared to traditional methods. The prefabricated nature of SIPs also minimizes on-site waste and promotes sustainable building practices. Despite their advantages, SIPs require careful handling to prevent damage to the insulating core during transportation and installation. Proper sealing of joints is also essential to maintain the panels' thermal performance and prevent air leakage.

2. **Light Concrete Building Systems**
   Light concrete building systems use lightweight concrete, such as autoclaved aerated concrete (AAC) or other lightweight aggregate materials, to create walling solutions that offer several benefits over traditional concrete (Alabi and Mahachi, 2022). AAC, for example, is produced by mixing cement, lime, sand, and an aerating agent, resulting in a material that is lighter than conventional concrete but still strong enough for structural applications. Its cellular structure provides improved thermal and acoustic insulation, which can enhance indoor comfort and reduce energy consumption for heating and cooling.

   The reduced weight of light concrete systems makes them easier to handle and install on-site, which can lead to cost savings in terms of labour and transportation. The lightweight nature also makes these systems particularly suitable for projects where reducing the overall weight of the structure is critical, such as in multi-story buildings or renovations. Furthermore, light concrete systems can be shaped or cut to fit various design requirements, offering flexibility in architectural applications. However, these systems still require specialized construction techniques to ensure proper reinforcement and anchoring for stability.

3. **Polyvinyl Chloride (PVC) Building Systems**
   PVC building systems use PVC, a versatile plastic material, to create modular walling systems that are durable and resistant to various environmental factors. These systems typically consist of interlocking PVC shells or modules that are reinforced with concrete or other materials to enhance their structural performance. The interlocking design enables quick and easy assembly, making PVC building systems suitable for projects that demand rapid construction, such as emergency shelters or temporary structures.

   One of the main advantages of PVC building systems is their resistance to moisture, rot, and insects, which makes them ideal for use in humid or coastal environments. The material is also easy to clean and maintain, contributing to the structure's longevity. Furthermore, PVC systems can be designed to accommodate a range of finishes, allowing for aesthetic flexibility. However, the environmental impact of PVC production and disposal must be considered, as it is a synthetic material derived from petrochemicals. Incorporating recycled PVC or developing recycling programmes for end-of-life materials can help mitigate some of these concerns.

4. **Container Building Systems**
   Container building systems involve using repurposed shipping containers as modular building units. These steel containers provide a strong and durable framework that can be easily transported and stacked to create multi-level structures. The modular nature of container buildings enables flexible design options and allows for rapid assembly on-site, making them suitable

for projects with tight construction schedules or limited access, such as in remote locations.

These systems are aligned with sustainable construction practices by promoting the reuse of existing materials and reducing waste. Additionally, the robust nature of the containers makes them resistant to harsh weather conditions, providing a high level of protection for occupants. However, container building systems require careful planning and adaptation to ensure proper insulation, ventilation, and interior finishes, as the steel walls can conduct heat and cold. Retrofitting containers to meet local building codes and standards for safety and comfort is also necessary.

5. **Building Blocks**

   Building blocks cover a range of masonry units, including traditional clay bricks, concrete blocks, and newer types of blocks made from alternative materials such as recycled aggregates or compressed earth. These units are commonly used in construction for their strength, durability, and ease of availability. Building blocks are generally manufactured on-site or locally, often using local labour, which supports community employment and reduces transportation costs.

   The modularity of building blocks allows for flexibility in design and construction, with the ability to accommodate various shapes, sizes, and finishes. Moreover, using blocks made from sustainable or recycled materials can contribute to reducing the environmental footprint of the construction project. However, the quality of building blocks must be strictly controlled to ensure they meet structural requirements, and proper mortar application, where appropriate, is necessary to maintain stability. Additionally, the manual labour involved in block construction can make the process more time-consuming than prefabricated systems.

6. **3D Construction Printing**

   3D construction printing is a cutting-edge technology that uses automated machinery to "print" building components layer by layer, typically using a concrete or cement-based material. The process can significantly reduce construction time and labour costs while allowing for complex and customized designs that are difficult to achieve with traditional methods. The automated nature of 3D printing also enhances precision, leading to less material waste and greater sustainability.

Although 3D construction printing shows promise for affordable and rapid construction, particularly in the housing sector, its widespread adoption in South Africa is still in its early stages. Agrément SA has not yet approved any 3D construction printing systems for general use, as comprehensive performance assessments are required to ensure safety, durability, and compliance with national building standards. Challenges such as material consistency, structural integrity, and regulatory acceptance must be addressed before 3D-construction printed buildings can become a mainstream solution.

The diversity of walling systems certified by Agrément SA highlights the potential for innovation in the construction industry. Each system offers unique benefits and addresses different challenges, contributing to a broader spectrum of building solutions that cater to various environmental, economic, and social needs in South Africa.

The classification of walling building systems, as outlined in Table 10.1 (based on the data available at the time), provides a broad overview of their performance and characteristics. While the list is comprehensive, further detailed analysis could enhance understanding. This aspect will be explored in subsequent sections, which will examine the building occupancies closely linked to these systems.

ANALYSIS OF THE BUILDING SYSTEMS

Occupancy Classification

The NBR (1977) classify buildings according to their intended purpose, covering thirty categories that include residential, commercial, healthcare facilities, educational institutions, garages, and storage spaces. Agrément certification aligns with these classifications, evaluating a product's performance in relation to specific occupancy requirements, as summarized in Table 10.1. An analysis of the distribution of occupancy classifications was conducted, with the results illustrated in Figure 10.2.

Insights from the data are summarized as follows:

1. **Limited Coverage for High-Risk Occupancies**
   The data reveals a significant gap in the certification of walling systems for high-risk occupancy classes. These classes include categories such as entertainment and public assembly (A1), theatrical and indoor sport (A2), outdoor sport (A5), high-risk commercial service (B1), exhibition hall (C1), hospital (E2), health care (E4), and high-risk storage (J1). Walling systems in these categories are subjected to stringent compliance requirements because any failure could pose serious risks to health and safety. For instance, buildings used for public gatherings or hospitals must meet higher standards for fire resistance, structural stability, and durability because they involve large numbers of people and potentially life-saving activities.

   The scarcity of certified systems for these occupancies may be due to the higher technical demands and more rigorous testing needed to ensure that the walling systems meet the required safety standards. In public assembly areas, such as theatres or indoor sports venues, structures must be designed to support large, moving crowds while minimizing the risk of collapse. Similarly, hospitals and healthcare facilities demand construction systems that can ensure structural integrity while providing essential acoustics and temperature control. The lack of certified walling systems suitable for these high-risk environments indicates a need for greater innovation and investment in advanced materials and technologies that can satisfy these stringent requirements. Addressing this gap could help improve safety in critical infrastructure, especially in emergency situations where reliable performance is non-negotiable.

TABLE 10.1
Building walling system stratification groups

Building groups	Names of building systems	Building occupancies*
1. Sandwich insulated panels	a. Spaceframe 2000 Building System	A3, D2, G1, H2, H3, H4
	b. Amsa Protea Building System	H3, H4
	c. FSM Building System	A3, H2, H4
	d. Kwikspacc Modular Building System	A3, D2, D3, G1
	e. MIB Building System	A3, B2, B3, D2, D3, E3, F2, G1, H3, H4
	f. SARDA Building System	A3, B2, B3, D2, D3, F2, G1, H2, H3, H4
2. Light concrete building systems	a. UCO Solidwall Building System	A3, B2, B3, E3, F2, G1, H3, H4, J2, J3
	b. Robust Building System	A3, B2, B3, D2, D3, E1, E3, F1, F2, F3, G1, H2, H3, H4
	c. Tilt-Up Prefabricated Building System	H3, H4
	d. Uvuyo Building System	B2, B3, F2, H3, H4
3. PVC building systems	a. Flex Building System	A3, B2, B3, F1, F2, F3, G1, H3, H4, J2, J3
	b. Luxwood Wall Panel Building System	B2, B3, D2, D3, F1, F2, F3, G1, H3, H4
	c. GHS Wall Technology Building System	A3, A4, B2, B3, C2, D2, D3, D4, E3, F2, F3, G1, H2, H3, H4, H5, J2, J3, J4
4. Container building systems	a. ITAS Modular Building System	G1
	b. Xtraspace Container Building System	A3, B2, B3, D1, D2, D3, F2, G1, H1, H2, H3, H4
5. Building blocks	a. Mega Building System	A3, B3, F2, G1, H3, H4
	b. Stumbelblock Building System	A3, B2, B3, D2, D3, F2, G1, H2, H3, H4
	c. Benex Masonry Building System	A3, A4, B3, D2, D3, F2, G1, H2, H3, H4
	d. Compressed Earth Building System	A3, A4, B2, B3, F2, G1, H2, H3, H4
	e. Hydraform Building System	A3, B2, B3, D2, D3, F2, G1, H2, H3, H4, J1, J2
	f. Ikhaya Brick Building System	H3, H4
	g. Izoblock Building System	A3, A4, B2, B3, D2, D3, E3, F1, F2, F3, G1, H1, H2, H3, H4
	h. Klevabrick Building System	A3, B2, B3, D2, D3, E3, F1, F2, F3, G1, H1
	i. Automapolyblock Building System	A3, B2, B3, D2, D3, F2, G1, H2, H3, H4

Note: *As per National Building Regulations Nomenclature and described in more detail in the next section.

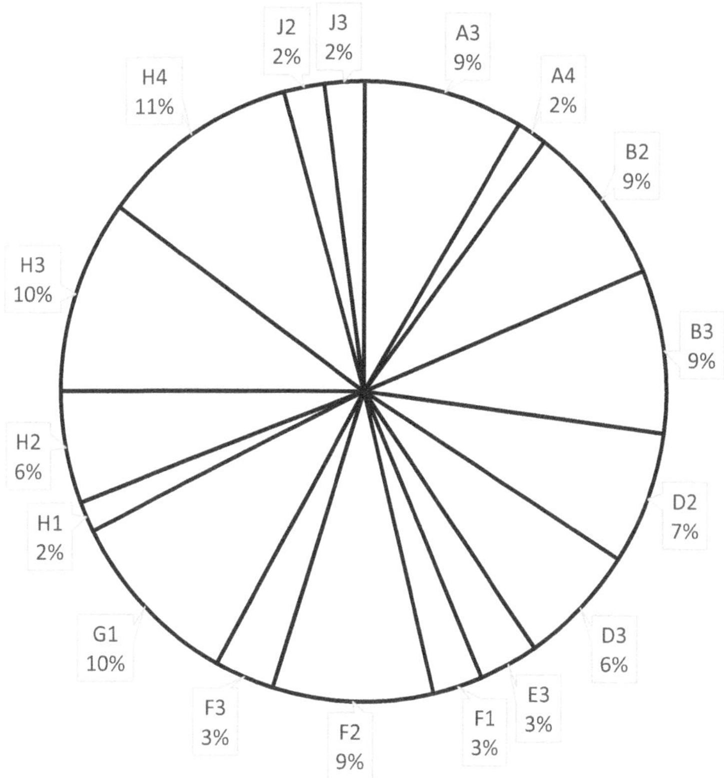

FIGURE 10.2 Distribution of occupancy classes (classes are as per NBR, 1977).

2. **Focus on Low-Risk Occupancies**
The data shows that the bulk of certifications are concentrated in low-risk occupancy classes, such as places of instruction (A3), low-risk commercial services (B3), moderate-risk industrial (D2), low-risk industrial (D3), small shops (F2), offices (G1), domestic residences (H3), and dwelling houses (H4). This trend suggests that innovative walling systems are more readily adopted in environments where the risks associated with building performance are lower. For example, office spaces and small retail shops may not have the same stringent requirements for fire resistance or structural resilience as hospitals or theatres, making it easier for new technologies to gain certification in these settings.

However, the high number of certifications in the H3 classification (multiple units on a site) presents specific challenges, particularly in low-income housing projects. In these developments, structural issues such as cracking, fire risks, and poor workmanship are prevalent, often exacerbated by limited budgets. The choice of walling system becomes critical in ensuring that housing for low-income communities is not only affordable but also safe and durable. Lightweight construction methods or alternative materials might offer cost savings, but without rigorous testing and certification, they may fail to provide the long-term reliability needed in these applications. Thus, while the emphasis on low-risk occupancies facilitates easier market entry for innovative systems, there is a need for additional measures to ensure that these solutions are also appropriate for settings where quality and safety are of paramount concern.

3. **Challenges and Opportunities**

   Adopting innovative building systems faces various challenges but also presents significant opportunities for growth and improvement. One of the major barriers to systems such as SIPs is the high initial capital investment required for manufacturing. Setting up a production line for SIPs demands substantial funding, which can be a deterrent for smaller companies or emerging markets. The high costs associated with the equipment and materials needed to produce SIPs may also limit their adoption to high-end markets, leaving out low-income or public sector applications where affordability is a key consideration.

   Additionally, SIPs and other modern systems can face resistance due to social and cultural factors, such as a preference for traditional construction aesthetics. In some communities, the appearance of innovative building materials may not align with established norms or expectations, making it harder for these systems to gain acceptance. Conversely, traditional block building systems are more familiar and widely accepted in many regions because of their solid appearance and established reputation. However, their slower construction speed compared to SIPs can be a disadvantage, particularly when rapid deployment is necessary, such as in disaster relief or emergency housing.

   The need for quick and efficient construction becomes especially evident during natural disasters or other crises, where the ability to rapidly deploy social infrastructure such as temporary shelters, clinics, or schools is crucial. In such cases, systems like SIPs and container building systems offer significant advantages. Their modular design allows for off-site manufacturing and quick on-site assembly, providing a solution that meets urgent demands for infrastructure while also promoting sustainability using prefabricated components. The challenge lies in scaling these technologies to make them more accessible for widespread use, particularly in regions with limited resources.

4. **Specialized Occupancy Classes**

   The certification data indicates limited coverage for specialized occupancy classes, which have unique requirements for construction materials and techniques. Only one system from the sample caters specifically to the E1 classification (place of detention), such as prisons or correctional facilities. This type of occupancy requires construction materials that offer robust structural strength and stability to prevent escapes, as well as enhanced acoustics and energy efficiency to maintain a controlled and safe environment for inmates and staff. Meeting these requirements involves using materials that can withstand significant impact and resist tampering or unauthorized access, making it a highly specialized market.

Another system aligns with the H1 classification for hotels, which introduces distinct demands such as ensuring guest comfort, safety, and structural integrity. Using steel containers for hotel construction provides a unique solution by offering modular and flexible building options, but it also necessitates additional scrutiny to ensure compliance with structural and fire safety standards. Hotels require high levels of insulation, soundproofing, and ventilation to meet guest expectations while also adhering to local building codes. Adapting steel containers to meet these standards poses both a challenge and an opportunity for innovative design approaches.

Therefore, the insights from the certification data underscore the importance of targeted research and development in walling systems to bridge gaps in high-risk occupancy classes, support low-income housing solutions, and capitalize on opportunities for rapid deployment and specialized applications. Expanding the range of certified walling systems across diverse occupancy categories can significantly enhance the safety, efficiency, and sustainability of the built environment in South Africa.

Energy and Thermal Performance

Evaluating energy and thermal performance is a crucial aspect of building construction, given the significant impact of buildings on overall energy consumption and environmental sustainability. This section provides an in-depth analysis of how walling systems contribute to energy efficiency, specifically in terms of heating, cooling, and maintaining indoor air comfort. In both residential and commercial buildings, these factors play a central role in determining energy usage and related costs. Efficient energy performance can help minimize the need for mechanical heating and cooling, thereby reducing electricity bills and lowering the carbon footprint of buildings.

Agrément SA assesses the energy efficiency of walling systems through rigorous testing using building energy simulation programmes. These programmes analyse the heating and cooling requirements of buildings under different conditions, such as during cold winter months and hot summer periods. As a benchmark, Agrément uses a standard brick house (SBH), which serves as a reference model for evaluating how various innovative walling systems perform relative to traditional brick construction. The SBH represents a typical building envelope commonly found in South Africa, offering a baseline for comparison that ensures consistency in the assessment of alternative construction systems.

Walling systems that match or surpass the energy efficiency of the SBH demonstrate their potential to reduce energy consumption in real-world applications. Many innovative walling products, especially those incorporating advanced materials or design features, have been shown to outperform the SBH in terms of energy usage. When additional insulation is included, these systems can significantly improve the building's thermal performance by reducing heat loss during winter and limiting heat gain during summer. The result is a more stable indoor temperature, which reduces the need for energy-intensive heating and cooling systems and enhances occupant comfort.

The study found that several walling systems exhibit outstanding energy efficiency, especially those within the SIPs, AAC, and lightweight concrete categories. Each system has unique characteristics that contribute to its superior thermal performance.

Using energy-efficient walling systems has far-reaching implications for sustainable construction, particularly in regions like South Africa, where energy scarcity and rising energy costs are major concerns. By lowering the energy demand for heating and cooling, these systems reduce greenhouse gas emissions associated with electricity generation, which is predominantly coal-based in South Africa. The reduction in energy consumption not only benefits the environment but also offers economic advantages for building owners and occupants, leading to significant cost savings over the building's lifespan.

Incorporating these innovative walling technologies into mainstream construction practices supports national and global goals for sustainable development. It aligns with policies aimed at reducing energy consumption in buildings and promoting using materials that minimize environmental impact. For example, the South African National Building Regulations (SANS 10400-XA, 2012) emphasize energy efficiency in buildings, and the adoption of high-performance walling systems can help developers meet these requirements more easily.

As awareness of energy efficiency and environmental sustainability continues to grow, there is increasing potential for the wider adoption of high-performance walling systems. However, several challenges must be addressed to facilitate this transition. For instance, there is a need for further education and training within the construction industry to ensure that architects, engineers, builders, and regulators understand the benefits and proper application of these systems. Additionally, government incentives and policies could be strengthened to encourage the use of energy-efficient building materials, such as through rebates for projects that achieve high levels of thermal performance.

Thus, energy-efficient walling systems such as SIPs, AAC, and lightweight concrete offer substantial benefits in reducing energy usage in buildings. Their adoption can play a significant role in achieving sustainable construction goals, particularly in energy-scarce regions where the cost of energy is a barrier to economic development. As the construction industry evolves, these innovative materials and systems will likely become more prevalent, driving a shift towards more sustainable building practices.

GENERAL DISCUSSIONS

This discussion explores the transformative potential of innovative building systems, focusing on modular construction and 3D construction printing as catalysts for sustainable urbanization in South Africa's affordable housing sector. Given the pressing need for affordable, quality housing in rapidly urbanizing regions, these modern construction methods present a significant opportunity to reshape the industry by enhancing the speed, cost-effectiveness, and environmental sustainability of housing delivery. This chapter has examined various walling systems certified by Agrément SA, including SIPs, light concrete building systems, PVC building systems, container building systems, building blocks, and emerging technologies such as 3D construction printing. Each system's performance attributes have been evaluated, revealing their potential advantages and limitations in addressing the housing challenges facing the country.

The stratification of walling systems based on structural and energy performance provides valuable insights into their applicability across different occupancy classes. While some systems excel in low-risk settings, others show promise for specialized applications, such as social infrastructure, in disaster response scenarios. SIPs, for example, offer superior energy efficiency and rapid deployment capabilities, making them suitable for emergency housing solutions. On the other hand, block building systems are favoured for their familiarity and robustness but may lag behind in terms of construction speed and thermal performance. These findings emphasize the importance of selecting appropriate construction systems based on specific project requirements, performance characteristics, and contextual challenges.

One of the significant challenges identified in the study is the limited coverage of high-risk occupancy classes. Certified walling systems are less prevalent in categories such as entertainment venues, hospitals, and high-risk storage facilities, where stringent safety and performance standards are critical. This scarcity suggests a gap in the certification and development of systems that can meet the demands of these high-risk settings, highlighting an area for further research and innovation. Conversely, there is a stronger presence of certified systems in low-risk occupancy classes, such as domestic residences and small offices, indicating that innovation has predominantly targeted these markets. However, even in these low-risk categories, challenges persist, particularly in low-income housing projects where factors such as budget constraints, structural concerns, and fire safety need to be carefully managed.

Engagement with system innovators has highlighted the practical hurdles in adopting modern construction methods. For instance, SIP products require substantial initial investment for manufacturing infrastructure, which can be a barrier to entry for smaller companies or projects with limited funding. Additionally, the aesthetic differences between traditional and modern methods may lead to social resistance, especially in communities where conventional brick-and-mortar construction is the norm. Similarly, while 3D construction printing holds the promise of revolutionizing the construction industry with its speed, cost savings, and reduced material waste, its adoption has been slow due to the need for comprehensive performance assessments and the lack of established regulations and standards.

The discussion also underscores the critical role of government support in promoting the uptake of modern construction practices. While past resolutions have favoured the use of MMC, there has been limited follow-through in terms of incentives, policy frameworks, and financial assistance to facilitate the widespread adoption of these systems. For sustainable urbanization to be realized, active government involvement is essential to create an enabling environment that encourages innovation, supports research and development, and provides clear pathways for certification and compliance. In particular, Agrément SA's certification process plays a vital role in validating the performance of innovative building systems, ensuring they meet stringent safety and quality standards. However, to accelerate the adoption of MMCs, it is necessary for the certification process to be complemented by policies that incentivize innovation and reduce the barriers to market entry.

This chapter advocates for an integrated delivery framework to address these challenges and foster the acceptance of modern construction methods. This framework would involve a multi-stakeholder approach that brings together system innovators, industry experts, regulatory bodies, and government agencies to collaborate on the development and implementation of innovative solutions. Key components of this framework would include comprehensive research on the performance and environmental impact of various systems, transparent decision-making processes informed by data, and the establishment of pilot projects to demonstrate the practical benefits of new technologies. The final step in this integrated approach would be Agrément SA's certification, ensuring that only rigorously tested and proven systems are approved for use, thereby mitigating risks and building confidence in the market.

Moreover, the emphasis on energy and thermal performance throughout the study reflects the growing importance of sustainability in construction. The potential of innovative walling systems to reduce energy consumption and improve indoor air quality presents significant advantages for regions like South Africa, where energy resources are scarce and costs are high. High-performance systems such as SIPs, AAC, and lightweight concrete enable the construction industry to contribute to national energy efficiency targets and support the transition towards greener building practices. Adopting these systems aligns not only with environmental goals but also carries social benefits, as it can improve living conditions for low-income households by reducing utility costs and enhancing thermal comfort.

CONCLUSION AND RECOMMENDATIONS

The testing, assessment, and certification of building products are pivotal in advancing innovation within the construction sector. A robust regulatory framework for MMCs in South Africa, reinforced by Agrément SA's impartial evaluations, ensures alignment with the NBR (1977). Although certificates issued by Agrément SA provide a solid foundation for industry stakeholders, a significant gap remains in translating research and certification into widespread practical application, particularly for innovative technologies like modular construction and 3D construction printing.

Bridging this gap requires a comprehensive approach to knowledge development that extends beyond stratified performance evaluations. It is crucial to explore the actual "as-built" performance of walling systems, considering local material availability, manufacturing methods, and societal acceptance. Building a holistic knowledge base that addresses these factors will facilitate the widespread adoption of sustainable construction products and practices. This will contribute to reducing carbon emissions and align the construction industry with national and global sustainability targets, supporting the transition towards greener building practices.

To unlock the potential of MMCs, South Africa's strategy must emphasize integrating sustainable construction methods, such as modular construction and 3D construction printing, which promise to lower carbon footprints through reduced material waste and energy consumption. The construction sector's alignment with sustainable urbanization goals will not only drive economic growth but also address social challenges by delivering affordable, energy-efficient housing. A collaborative effort to develop and utilize an expansive knowledge base will enable the industry to meet carbon reduction targets and propel the nation towards sustainable development.

REFERENCES

Agrément SA. (2010). The performance concept. www.agrement.co.za, Pretoria. Agrément South Africa.

Agrément South Africa Act (Act 11 of 2015) (20 October 2017). Government Gazette 41186, Pretoria, South Africa.

Alabi, S. and Mahachi, J. (2022). Mechanical properties of sustainable concrete made with ceramic and sandcrete block wastes. *Materials Today Proceedings*. 62, pp. S44–S48.

Burger, S. (2014). *Innovative building technologies can aid delivery of social infrastructure*. Creamer Media. South Africa

Buswell, R. A., Soar, R. C., Gibb, A. G., & Thorpe, T. (2018). Digital concrete: opportunities and challenges. *The European Physical Journal Special Topics*, 227(22), 2641–2652.

CIB (2003). *Performance-based building: First international state-of-the-art report*. CIB Development Foundation: PeBBu Thematic Network, Rotterdam.

Fairclough, J. (2002). *Rethinking construction innovation and research: a review of government R&D policies and practices*. Department of Trade and Industry/Department of Transport, Local Government, London.

Foliente, G. (2000). Developments in performance-based building codes and standards. *Forest Products Journal*, 50(7/8), pp. 12–21.

Gibb, A., & Shanks, G. (2019). Toward a new theory of construction as production. *Automation in Construction*, 98, 4–16.

Haron, A. T., Ismail, N., Mohamad, N., Zainun, N. Y., & Ismail, M. (2020). Barriers to the implementation of industrialized building system (IBS) in Malaysia. *Sustainability*, 12(10), 4068.

Hartkopf, V., Loftness, V. and Mill, P. (2008). The concept of total building performance and building diagnostics. *Building performance: Function, preservation and rehabilitation, ASTM STP 901*, G. Davise, Ed., American Society for Testing and Materials, Philadelphia, 1986, pp. 5–22.

Kaneko, Y., Tomita, K., Kondo, Y., & Inamura, H. (2018). Innovation management strategies for the Japanese construction industry. *Procedia CIRP*, 70, 524–529.

Khoshnevis, B., Dutton, R., Kolb, J. F., Mankins, J. C., Todd, M., & Vakhshouri, B. (2016). Automated construction by contour crafting—Related robotics and information technologies. *Automation in Construction*, 68, 21–33.

Li, Z., Zhang, Y., Shen, G. Q., Yu, A. T., & Zhang, W. (2020). A review of the application of prefabrication in China's construction industry. *Automation in Construction*, 113, 103091.

Ling, F. Y. Y., Tse, Y. K., Wang, Y., Shen, L., & Zhang, M. (2020). Critical success factors for offsite manufacturing construction in European countries. *Journal of Cleaner Production*, 271, 123027.

Love, P. E., Edwards, D. J., & Goh, Y. M. (2019). Modelling the readiness of the construction industry for off-site manufacturing. *Construction Innovation*, 19(4), 484–508.

Lützkendorf, T., & Balouktsi, M. (2017). Innovations in building retrofitting: The potential of building typologies. *Energy Policy*, 101, 381–394.

Ma, T., Lu, W., Skitmore, M., & Zhang, X. (2018). Challenges and potential solutions for prefabricated and modular construction: A case study of Australia. *Journal of Cleaner Production*, 180, 401–415.

Mahachi, J. (2021). Development of a construction quality assessment tool for houses in South Africa. *Acta Structilia*. 28(1), pp.91–116.

NBR (1977). National Building Regulations and Building Standards Act. No. 103 of 1977, as amended. Government Gazette. Pretoria. South Africa.

Ogunlana, S., & Papadonikolaki, E. (2018). A framework for the adoption of off-site manufacturing in the USA. *Journal of Construction Engineering and Management*, 144(1), 04017110.

Pan, W., Zhang, Q., Liu, J., & Wu, P. (2018). Barriers to the adoption of off-site construction in China: An empirical study. Journal of Cleaner Production, 195, 97–106.

SANS 10400 (2012): The Application of the National Building Regulations, SABS, Pretoria, Republic of South Africa.

Wong, T. N., Wong, K. A., & Ho, S. P. (2019). Barriers to the adoption of building information modelling in the Hong Kong construction industry. *Journal of Professional Issues in Engineering Education and Practice*, 145(3), 04019006.

Yuan, Y., Wang, Y., Gu, Y., & Ding, Z. (2018). A review of the housing industrialization policies in China: The past, present, and future. *Journal of Cleaner Production*, 176, 787–801.

11 3D Printing for Inclusive Health Care

Insights from a Delphi Study

Beata Mrugalska and Tihomir Dovramadjiev

INTRODUCTION

In recent years, the entire world has recognized the importance of healthcare systems and the consequences that arise while deprioritizing human life for other considerations. It became evident that health care needs to be inclusive, ensuring that everyone can access medical treatment during emergencies and for preventive care (Filho et al., 2021; Mrugalska & Dovramadjiev, 2023). Medical systems should cater to every individual, including those facing unique situations, and offer specialized treatment when contemporary technology is necessary. It includes addressing medical issues that could be partially or completely resolved through ergonomic and bioengineering advancements (Dovramadjiev et al., 2024). According to the United Nations Convention on the Rights of Persons with Disabilities, people with disabilities have the right to participate in society on an equal basis with others. It regards people with physical, mental, intellectual, cognitive, or sensory challenges that make them have difficulties when interacting with numerous obstacles, which might impede their full and equal involvement in social life compared to others (Degener, 2016; Yang, 2024). Following the report of the Academic Network of European Disability Experts, it is required to (Waddington & Priestley, 2021):

- Provide essential health services tailored for individuals with disabilities due to their unique needs, including early diagnosis and interventions, as well as services aimed at reducing and preventing further disabilities.
- Require healthcare professionals to deliver care of equivalent quality to persons with disabilities as they offer to others.

However, in reality, few countries provide sufficient quality services for people with special needs. Unfortunately, this happens, often due to the limited abilities and possibilities of medical and technical personnel (particularly in the case of medical implants) (Chakravarty, 2022; Dovramadjiev et al., 2023). Therefore, it is essential to develop a novel interdisciplinary curriculum encompassing such topics as ergonomics and bioengineering, addressing the needs of individuals with disabilities to foster a healthcare system. This initiative will improve inclusion, equality, and diversity as it can enhance the knowledge and skills of designers and producers of 3D medical

DOI: 10.1201/9781003625360-11

implants. Additionally, the ergonomic and bioengineering sector is undergoing a significant shift, thanks to the introduction of advanced software for implant design and the possibility of printing them using 3D printers. The impact of 3D printing on the medical sector is transformative, with digital skills serving as the key driver of innovation in implants.

This chapter aims to investigate which knowledge and skills should be transferred to both students (to enhance their job perspective) and higher education teachers (to increase their influence on students from the fields of medicine, ergonomics, industrial design, mechanical engineering, bioengineering, and materials science and engineering). To accomplish this, we suggest analyzing these sectors' current state and needs for recruits, illustrating that crucial hard skills pertain to design expertise and proficiency with relevant software. In order to achieve it, we performed a Delphi study on national and international levels. It allowed us to differentiate broad meta-levels of education in 3D printing for inclusive health care.

LITERATURE REVIEW

3D printing, also known as the additive technique or manufacturing, has significantly transformed the healthcare industry. It enables the creation of customized medical devices, implants, anatomical models, and drug delivery and pharmaceutical applications (Wang et al., 2021). It is mainly used for the creation of patient-specific prosthetic limbs and orthotic devices to enhance their fit, comfort, and functionality. It also finds its applications in crowns, bridges, and implants, which can be tailored to individual patients, improving dental care outcomes (Mrugalska et al., 2021; Pavlova et al., 2023; Pavlova et al., 2024). In surgical planning and training, 3D-printed models of organs and tissues help surgeons plan and practice complex surgeries, decreasing operative time and improving outcomes. These models are also valuable in medical education, providing hands-on learning opportunities for students and professionals (Barrios-Muriel et al., 2020). Furthermore, 3D printing enables the making of personalized drug delivery devices and tablets with specific dosages and release characteristics (Tan et al., 2020). Thanks to it, custom surgical tools and guides can be produced to improve precision during complex procedures. Moreover, bioprinting, which is an emerging field, involves printing living cells to create tissues and potential organs for transplantation (Matai et al., 2020).

3D printing creates a 3D object using computer software by layer-by-layer depositing material on a substrate. Two fundamental processes are involved in all 3D printing methods. The first involves designing the object using computer software, and the second involves using a 3D printer to deposit or create the object. A variety of computer programs, including AutoCAD (computer-aided design), 3D Slash, SketchUp, Fusion 360, and Solidworks, can be used to design 3D objects. Using slicer software such as KISSlicer, Slic3r, OctoPrint, Simplify3D, and Cura, we cut the created design. A 3D design file (STL file) can be transformed into G-code that can be read by a 3D printer using this 3D slicer software (Mohammed et al., 2021).

In the next step, the formed design is sliced through slicer software such as KISSlicer, Slic3r, OctoPrint, Simplify3D, and Cura. The software converts a 3D design file (STL file) to a 3D printer-readable G-code (Virmani et al., 2024). The type of 3D printer determines the object deposition procedure. Generally, on the basis of the method of material layering, 3D printers work using inkjet, extrusion, or laser-based systems.

Inkjet printing is categorized into continuous inkjet (CIJ) and drop-on-demand (DOD) types. CIJ uses a continuous ink flow, controlled via piezoelectric transducers and electrostatic fields, whereas DOD inkjet printing relies on droplet formation only when needed, utilizing less ink and offering two print head types: thermal and piezoelectric. Piezoelectric print heads are preferred for room temperature applications, as thermal print heads risk degrading active components. DOD printing splits into two methods: drop-on-liquid, which builds microstructures layer-by-layer for high-capacity loading, and drop-on-solid, used for controlled delivery systems, exemplified by the FDA-approved product Spritam®. Drop-on-solid printing deposits droplets onto powder beds, offering options for high drug loading with application programming interfaces (API) and controlled release using various polymers as binder ink. However, it faces challenges like product brittleness, stability issues, and potential nozzle blockages in ink preparation (Shah et al., 2021; Zapka, 2022).

Extrusion-based 3D printing techniques focus on two methods: pressure-assisted microsyringe (PAM) and fused deposition modeling (FDM). The PAM technique involves using semisolid materials or pastes to create 3D structures, emphasizing the need for material optimization and control over printing parameters like pressure and speed. It is suitable for heat-sensitive materials but has limitations like lower resolution in comparison to other methods. FDM, on the other hand, uses filaments made from melted polymers and provides good mechanical strength and printing resolution without the need for organic solvents. However, it is less suitable for thermosensitive active ingredients due to high processing temperatures. While FDM allows slow release and can adjust doses through in-fill patterns, PAM is better for high loading but requires careful solvent selection to prevent stability problems. While both techniques have unique advantages, such as high loading capacity for PAM and high-resolution printing for FDM, each has limitations that require careful consideration depending on the specific application (Mandava et al., 2022).

On the basis of liquid and powder solidification, we can differentiate two types of laser-based 3D printing techniques: stereolithography (SLA) and selective laser sintering (SLS). SLA uses UV laser beams to solidify photosensitive polymeric liquids into precise 3D structures, with high resolution up to 0.2 μm. However, it faces drawbacks such as a lack of FDA-approved resins, potential instability, and cytotoxicity issues. SLA can be used for creating loaded hydrogels and microstructures for delivery but is limited by the availability of suitable resins. SLS involves sintering a powder bed with a laser beam, allowing layer formation without melting, using no organic solvents. It offers a single-step process with high resolution but suffers from potential API degradation and limitations in material choice and hollow structure printing. Both techniques present unique advantages and challenges based on material properties and the process dynamics used, such as laser intensity and exposure duration in SLA and laser focus and particle size in SLS (Song et al., 2024).

Despite its enormous advantages in the healthcare industry, 3D printing has several drawbacks. They primarily deal with technology, dosage form manufacturing, safety, quality assurance, regulatory considerations, and clinical pharmacy application. In practice, it appears that each type of 3D printer has some technological limitations. While heat and laser-based systems may experience deprivation of the used API, nozzle-based systems may have a blockage of the nozzle (Parhi, 2021). Furthermore, drug-excipient incompatibility is also a significant issue that needs to be resolved. Additionally, the final product may have surface and structural defects that must be fixed by optimizing some manufacturing factors. Another challenge can be the right choice of materials for 3D printing. It is often required that the materials must be biocompatible and biodegradable (Lamichhane et al., 2019). Also, they are not allowed to generate toxic substances, which can be respiratory or skin irritants, while processing such as heating, extrusion, or fusion of certain materials. Thus, it is required to follow adequate safety measures and standard operating guidelines (Fatouros, 2018). Furthermore, the challenge can be to employ highly skilled technical operators to cope with the need to provide high-quality printed dosage forms and techniques (Tren et al., 2018). From a financial point of view, in health care, a 3D printer should be fast, easy to use, and economical and have a good resolution. Since now, there are some FDA regulatory frameworks. For example, in 2017, the FDA issued the regulatory requirements for the manufacturing of medical devices (FDA, 2017). "Good Manufacturing Practice" (GMP) facility certification should be required for any location using a 3D printer to manufacture and dispense pharmaceutical items (23). However, there is still a lack of regulations and measures to protect against counterfeit medications. According to the World Health Organization, substandard or falsified medicines affect 10.5% of low- and middle-income countries due to inadequate pharmaceutical governance, insufficient technical capabilities, and ineffective supply chain management (Trenfield et al., 2019). This matter not only has substantial health implications but also poses a direct threat to global health security and the efforts to fulfill the United Nations Sustainable Development Goal 3.8, which aims to provide universal access to safe and effective essential medicines (WHO, 2017).

METHODS

The Delphi method was used to perform this study as it is suitable for developing a variety of potential answers, investigating underlying presumptions that could result in disparate conclusions, gathering data that could lead to agreement, and correlating well-informed conclusions. The essential element of its process is participants' anonymity when voicing their opinions. Delphi alleviates issues that can arise from a small number of influential or privileged people controlling the group. It also avoids the bandwagon tendency, susceptibility to manipulation, and people's hesitancy to change their thoughts in front of others, which are the drawbacks of in-person conversations (Linstone & Turoff, 2011). In our study, we followed a careful process resulting in the choice of criteria for selecting national and international experts. The experts were selected based on their educational background, experience, and employment. They were representatives of three groups: academia, practice, and students. It was assumed that at least one of them was from each group. The expert had to meet the following three criteria:

- Educational background (university degree).
- Period practice – at least 1 year of using qualifications.
- Field of expertise related to the project – engineering, ergonomics, medicine, physiotherapy, software engineering, IT, education, psychology, technical device operation, health care, rehabilitation, and dentistry.

In a case of students, they were supposed to be at the master's level in the field of study related to the project, the same as academicians and practitioners. We also assumed that we would invite people who have experience in working with people with special needs (individuals who require assistance for disabilities that may be related to medical, mental, or psychological issues).

The opinions were collected from a group of experts through a questionnaire on Google Form, which consisted of three parts:

- Part A: Theory and practice
- Part B: Technologies
- Part C: Teaching and training methodology.

The first part included 58 statements, the second one contained 41 statements, whereas in the last part there were 6 statements. Each of these statements was assessed separately using a scale:

- No answer
- Strongly disagree
- Disagree
- Neither agree nor disagree
- Agree
- Strongly agree.

The first Google Form was available for almost five weeks, from November 25, 2022, to January 1, 2023, whereas the second round was available for seven days from 6th January 2023. We contacted the experts by sending the link to the questionnaire via email, and they responded anonymously. Each participant was asked to give estimates and assumptions to the Delphi facilitator (the corresponding author), who collected the information via Google Form and prepared a first summary report. Helped by the facilitator, the invited respondents reviewed the summary report, giving updated information to the facilitator, who reviewed the material and issued a second report. This process continued until all participants reached a consensus. We assumed that the consensus was achieved when 60% or more answers were "agree" or "strongly agree" for a statement.

TABLE 11.1
List of experts in the first round of the Delphi method

No	Country	Expertise sector	Number of experts	Number of experts working with people with special needs
1	Bulgaria	Practitioner	9	2
2	Bulgaria	Academia	2	0
3	Bulgaria	Student	9	1
4	Poland	Practitioner	4	3
5	Poland	Academia	5	3
6	Poland	Student	1	1

TABLE 11.2
List of experts in the second round of the Delphi method

No	Country	Expertise sector	Number of experts	Number of experts working with people with special needs
1	Estonia	Academia (engineering, ergonomics)	1	1
2	Malaysia	Academia	2	0
3	Turkey	Practitioner, academia (education, IT, software engineering)	3	1
4	Latvia	Academia (ergonomics)	1	1
5	Romania	Academia	2	0

The Delphi study was conducted on two levels:

- National level
- International level

in two countries (Poland and Bulgaria) (Table 11.1) and then in an international Delphi study (Table 11.2).

In the first round, 30 participants (20 – Bulgaria and 10 – Poland) answered the questionnaire. They represented two sectors: academia (teachers and students) and practice (IT, engineering, dentistry, education, health care, medicine, software engineering, technical device operation, engineering, ergonomics, and medicine).

In the second round, we invited nine international experts from such countries as Estonia, Malaysia, Turkey, Latvia, and Romania. However, this time, statements were assessed, and the responses were aggregated and shared with them after the first round. Again, we assumed that the consensus was achieved when 60% or more answers were "agree" or "strongly agree" for a statement. In a case where the decision of our respondents was lower than 60%, the statement was rejected and was not included as an idea for a topic of the course. For further analysis, all accepted statements were investigated using statistics. We calculated the mean and standard deviation (SD) to identify which and how much the topics in the statements were vital in teaching in our potential course. It results from the fact that a SD measures how dispersed the data are to the mean. Low SD means data are clustered around the mean, and high SD indicates data are more spread out.

RESULTS AND DISCUSSION

The following three research questions were addressed in this study:

Question 1: What should be theoretical and practical topics in the scope of ergonomics and bioengineering in a new curriculum about 3D printing for inclusive health care (Part A)?

Question 2: What technologies should be taught in a new curriculum about 3D printing for inclusive health care (Part B)?

Question 3: What teaching and training methodology should be used in a new curriculum about 3D printing for inclusive health care (Part C)?

Table 11.3 presents the statements that achieved consensus in the first round and their results.

As it can be seen, the experts achieved consensus on 22 out of 58 statements from Part A, 12 out of 41 from Part B, and 3 out of 6 from Part C, indicating that they should be included in a new curriculum. In most cases, the mean was lower than 4.0, while the SD was about 1.0. It indicated the need to organize a second round of Delphi where international experts shared with us their opinions. In Table 11.4, we presented the results of the second round of Delphi.

As it can be noticed, the experts achieved consensus on 12 out of 58 statements from Part A, 8 out of 41 from Part B, and 3 out of 6 from Part C, indicating that they should be included in a new curriculum. They agreed or fully agreed that it is needed to present such topics as ergonomics and work environment for people with special needs and medical standards and rules of certification of medical products. According to them, contactless measurement techniques for anthropometric data gathering as well as orthosis/prosthesis modeling software are crucial technologies for teaching 3D printing. The exception was only noticed for medical imaging technologies where the mean was 3.89 and the SD was 1.05, indicating that the data observed are quite spread.

Based on a literature review on the topic and a qualitative analysis of the responses from the experts, three broad meta-levels of education in 3D printing for inclusive health care were derived:

- Macrolevel: theories and practices
- Mesolevel: technologies
- Microlevel: teaching and training methodologies.

Under these three levels, the content considered important by the experts can be categorized into areas briefly characterized in Table 11.5.

TABLE 11.3
The selected results of the first round of Delphi

No	Statement	%	Mean	SD
				Part A
1	Ergonomics: efficiency–safety–convenience in transforming and developing different environments, especially work environment adjusted to work with people with special needs	0.73	4.04	0.76
2	Fundamental topics of muscle anatomy, physiology, treatment, and muscle modeling	0.63	3.78	0.89
3	Anthropometry: human body data in ergonomic assessments and developments	0.60	3.81	0.96
5	Mechanical behavior of biological tissues and systems	0.60	3.64	0.87
7	Kinematics and kinetics of movement and muscle force system	0.63	3.86	0.80
8	Knowledge engineering and expert systems	0.63	4.04	0.77
11	Computational methods in biomedical technology using finite elements	0.63	3.74	0.94
12	Kinematics, dynamics, and energetic aspects of human movement using biomechanical models	0.70	3.89	0.83
13	Biomechanical models of the musculoskeletal system based on the anthropometry of the human body and the mechanical laws of movement	0.60	3.74	1.10
14	Biomechanical applications in human movement	0.60	3.79	1.03
15	Basic concepts on solid mechanics	0.67	3.81	0.96
17	Biomaterials in biomedical applications	0.60	3.78	0.97
23	Corrosion and degradation mechanisms of biomaterials in different applications	0.60	3.78	0.85
25	Principles of design and operation of medical equipment	0.63	3.81	0.74
26	Principles of design and operation of sensors	0.60	3.74	0.90
31	Medical device regulations	0.60	3.67	0.73
35	Human motor behavior simulation and virtualization; human body modeling; and application of captured human motion data to virtual models and scenes to simulate interaction	0.63	3.89	0.89
38	Intelligent (automated) computer-aided design models in medicine	0.63	3.78	1.01
40	Principles of design, material selection, and production of personalized implants and endoprostheses	0.67	3.82	0.98
42	Methodologies of design and production of personalized implants and endoprostheses based on medical imaging data and additive manufacturing technologies	0.60	3.75	0.97

52	Medical standards and rules of certification of medical products	0.73	3.89	1.03
57	Accessibility for people with special needs	0.60	3.64	0.87
				Part B
3	Orthosis/prosthesis modeling software	0.70	4.04	0.90
4	Software to the scan analysis of a healthy arm/leg of a patient with an arm/leg after injury	0.60	3.55	1.18
5	Software for the analysis of the mechanics of movement on the basis of real measurements	0.67	3.76	1.12
6	Operation, inspection, and maintenance of diagnostic and therapeutic equipment and software within the healthcare facility, including assistance to physicians during the examination of patients	0.60	3.78	1.09
8	Contactless measurement technologies for anthropometric data gathering (such as 3D scanning and photogrammetry)	0.70	4.04	0.84
10	Free and commercial software for medical imaging and digital mesh processing	0.63	3.88	0.95
27	CAx (CAD/CAM/CAE) software is a crucial part of the bioengineering course and should not be limited to just the basics	0.60	3.70	1.17
28	Basic courses in programming languages (C++, Python, Java, and others)	0.60	3.79	1.13
36	Software tools for visualization of medical data	0.70	4.04	0.81
39	Utilization of any source of medical 3D data to digitally design a personalized medical device (e.g., an implant or a prosthesis)	0.60	3.88	0.99
40	Technologies for measurement of motion kinematics: interfacing features; sensors; information recognition and extraction; signal processors; noise identification and suppression; offline and real-time tracking; and calibration and 3D reconstruction	0.70	3.85	0.77
41	Medical imaging technologies (such as computed tomography, magnetic resonance imaging, and ultrasound examination)	0.63	3.79	1.05
				Part C
4	Brainstorming and teamwork	0.70	3.74	0.86
5	Case studies, films, and presentations	0.63	3.68	0.94
6	Interactive lectures and discussions	0.80	4.03	1.02

TABLE 11.4
The selected results of the second round of Delphi

No	Statement	%	Mean	SD
				Part A
1	Ergonomics: efficiency–safety–convenience in transforming and developing different environments, especially work environment adjusted to work with people with special needs	1.00	4.78	0.44
2	Fundamental topics of muscle anatomy, physiology and treatment, and muscle modeling	0.89	4.00	0.50
7	Kinematics and kinetics of movement and muscle force system	0.89	4.33	0.71
8	Knowledge engineering and expert systems	0.89	4.33	0.71
12	Kinematics, dynamics, and energetic aspects of human movement using biomechanical models	0.78	3.89	0.93
13	Biomechanical models of the musculoskeletal system based on the anthropometry of the human body and the mechanical laws of movement	0.78	4.11	1.05
14	Biomechanical applications in human movement	0.89	4.22	0.97
17	Biomaterials in biomedical applications	0.89	4.00	0.87
35	Human motor behavior simulation and virtualization; human body modeling; and application of captured human motion data to virtual models and scenes to simulate interaction	0.89	4.44	0.73
38	Intelligent (automated) computer-aided design models in medicine			
40	Principles of design, material selection, and production of personalized implants and endoprostheses	0.89	4.11	0.60
52	Medical standards and rules of certification of medical products	1.00	4.56	0.53
				Part B
3	Orthosis/prosthesis modeling software	1.00	4.44	0.53
4	Software to the scan analysis of a healthy arm/leg of a patient with an arm/leg after injury	0.89	4.33	1.00
8	Contactless measurement techniques for anthropometric data gathering (such as 3D scanning and photogrammetry)	1.00	4.56	0.53
10	Free and commercial software for medical imaging and digital mesh processing	0.78	4.11	1.05
27	CAx (CAD/CAM/CAE) software	0.78	4.00	1.00
36	Software tools for visualization of medical data	0.78	4.00	1.00
40	Technologies for measurement of motion kinematics: interfacing features; sensors; information recognition and extraction; signal processors; noise identification and suppression; offline and real-time tracking; and calibration and 3D reconstruction	0.89	4.22	0.67
41	Medical imaging technologies (such as computed tomography, magnetic resonance imaging, and ultrasound examination)	0.67	3.89	1.05
				Part C
4	Brainstorming and teamwork	1.00	4.22	0.44
5	Case studies, films, and presentations	1.00	4.22	0.44
6	Interactive lectures and discussions	1.00	4.44	0.53

TABLE 11.5
Curriculum content proposal for the course in 3D printing for inclusive health care

Level	Content
Macro	Ergonomics: efficiency–safety–convenience in transforming, developing different environments, especially work environment adjusted to work with people with special needs; fundamental topics of muscle anatomy, physiology and treatment, and muscle modeling, kinematics and kinetics of movement and muscle force system, knowledge engineering and expert systems, kinematics, dynamics and energetic aspects of human movement using biomechanical models, biomechanical models of the musculoskeletal system based on the anthropometry of the human body and the mechanical laws of movement, biomechanical applications in human movement, biomaterials in biomedical applications, and human motor behavior simulation and virtualization; human body modeling; and application of captured human motion data to virtual models and scenes to simulate interaction, intelligent (automated) computer-aided design models in medicine, principles of design, material selection and production of personalized implants and endoprostheses, and medical standards and rules of certification of medical products
Meso	Contactless measurement techniques for anthropometric data gathering (such as 3D scanning and photogrammetry), free and commercial software for medical imaging and digital mesh processing, CAx (CAD/CAM/CAE) software, orthosis/prosthesis modeling software, software for the scan analysis of a healthy arm/leg of a patient with an arm/leg after injury, software for visualization of medical data, medical imaging technologies (such as computed tomography, magnetic resonance imaging, and ultrasound examination) technologies for the measurement of motion kinematics: interfacing features; sensors; information recognition and extraction; signal processors; noise identification and suppression; offline and real-time tracking; and calibration and 3D reconstruction
Micro	Brainstorming, case studies, discussions, films, interactive lectures, presentations, and teamwork

The study is limited to English-speaking 3D printing specialists with access to the Internet. Therefore, the results may not include the opinions of certain non-English speakers where the technical infrastructure does not allow access to the Internet. Due to the small sample size, complex statistical analysis could not be done, which means the results are descriptive, with the mean and SD being used to represent the experts' evaluations.

CONCLUSIONS

Among the various manufacturing processes, 3D printing is a unique additive method within the industry. It facilitates the production of three-dimensional solid items in nearly any shape, starting from a digital design. Initially perceived as an ambitious idea, recently medical 3D printing has also been made possible through significant time and financial commitments. However, there is still a need to enhance medical and technical staff to increase the production of 3D implants. To achieve it, we suggest to developing an interdisciplinary curriculum for higher education that encompasses various subjects, ranging from ergonomics to bioengineering. Such an initiative can enhance inclusion and diversity by refining the skills of those designing and producing 3D medical implants. Consequently, such a program could help in establishing a more inclusive environment that addresses the needs of the broader community, including those with special needs. Therefore, in our chapter, we proposed three broad meta-levels of education in 3D printing for inclusive health care. We defined them as macro-, meso-, and microlevels, and they respond to theories and practices, technologies, and teaching and training methodologies, respectively. They are based on our Delphi study, where we achieved consensus for 23 out of 105 statements after the second round. We proposed to include them in a new curriculum for 3D printing for inclusive health care.

REFERENCES

Barrios-Muriel, J., Romero-Sánchez, F., Alonso-Sánchez, F. J., & Salgado, D. R. (2020). Advances in orthotic and prosthetic manufacturing: A technology review. *Materials*, 13(2), 295. https://doi.org/10.3390/ma13020295

Chakravarty, S. (2022). Resource constrained innovation in a technology intensive sector: Frugal medical devices from manufacturing firms in South Africa. *Technovation*, 112, 102397. https://doi.org/10.1016/j.technovation.2021.102397

Degener, T. (2016). Disability in a human rights context. *Laws*, 5(3), 35. https://doi.org/10.3390/laws5030035

Dovramadjiev, T., Dimova, R., Dimov, D., & Manolova, P. (2024). Human-centered innovations in healthcare education: Digital skills for ergonomics and bioengineering advancement. In International Conference on Computer and Communication Engineering (pp. 209–216). Cham: Springer Nature Switzerland. https://doi.org/10.1007/978-3-031-71079-7_17

Dovramadjiev, T., Pavlova, D., Filchev, R., Dimova, R., Dimov, D., Kavaldzhieva, K., & Mrugalska, B. (2023). Technological and human approach in complete restoration of the dental teeth jaws with biocompatible materials. In International Conference on Smart Trends in Computing and Communications (pp. 497–506). Singapore: Springer Nature Singapore. https://doi.org/10.1007/978-981-99-0769-4_43

Fatouros, D. G. (2018). Recent advances in pharmaceutical dosage forms and devices using additive manufacturing technologies. *Drug Discovery Today*, 24(2), 636–643. https://doi.org/10.1016/j.drudis.2018.11.019

FDA (2017). Technical considerations for additive manufactured medical devices: Guidance for industry and Food and Drug Administration staff. *FDA*. www.fda.gov/regulatory-information/search-fda-guidance-documents/technical-considerations-additive-manufactured-medical-devices

Filho, S. Á., Ávila, J. S., Mrugalska, B., de Souza, N. F., Gomes de Carvalho, A. P. M., & Gonçalves, L. R. (2021). Decision making in health management during crisis: A case study based on epidemiological curves of China and Italy against COVID-19. *International Journal of Environmental Research and Public Health*, 18(15), 8078. https://doi.org/10.3390/ijerph18158078

Lamichhane, S., Bashyal, S., Keum, T., Noh, G., Seo, J. E., Bastola, R., et al. (2019). Complex formulations, simple techniques: Can 3D printing technology be the Midas touch in pharmaceutical industry? *Asian Journal of Pharmaceutical Sciences*, 14(4), 465–479. https://doi.org/10.1016/j.ajps.2019.04.001

Linstone, H. A., & Turoff, M. (2011). Delphi: A brief look backward and forward. *Technological Forecasting and Social Change*, 78(9), 1712–1719. https://doi.org/10.1016/j.techfore.2010.09.011.

Mandava, K., Thakur, S., Kadimcharla, K., & Katakam, P. Extrusion-Based 3D Printing Technology: A Revolution in Pharmaceutical Drug Manufacturing. In *Handbook of 3D Printing in Pharmaceutics*. Katakam, P., Kumar, R., Ranjan, N., & Babbar, A. (Eds.). (pp. 82–100). Boca Raton: CRC Press.

Matai, I., Kaur, G., Seyedsalehi, A., McClinton, A., & Laurencin, C. T. (2020). Progress in 3D bioprinting technology for tissue/organ regenerative engineering. *Biomaterials*, 226, 119536. https://doi.org/10.1016/j.biomaterials.2019.119536

Mohammed, A. A., Algahtani, M. S., Ahmad, M. Z., Ahmad, J., & Kotta, S. (2021). 3D Printing in medicine: Technology overview and drug delivery applications. *Annals of 3D Printed Medicine*, 4, 100037. https://doi.org/10.1016/j.stlm.2021.100037

Mrugalska, B., & Dovramadjiev, T. (2023). A human factors perspective on safety culture. *Human Systems Management*, 42(3), 299–304. https://doi.org/10.3233/HSM-220041

Mrugalska, B., Dovramadjiev, T., Pavlova, D., Filchev, R., Stoeva, M., Bozhikova, V., & Dimova, R. (2021). Open source systems and 3D computer design applicable in the dental medical engineering Industry 4.0 – sustainable concept. *Procedia Manufacturing*, 54, 296–301. https://doi.org/10.1016/j.promfg.2021.09.002

Parhi, R. (2021). A review of three-dimensional printing for pharmaceutical applications: Quality control, risk assessment and future perspectives. *Journal of Drug Delivery Science and Technology*, 64, 102571. https://doi.org/10.1016/j.jddst.2021.102571

Pavlova, D., Dovramadjiev, T., Daskalov, D., Dimova, R., & Mrugalska, B. (2023). File formats and digital data transfer for dental needs. In International Congress on Information and Communication Technology (pp. 483–495). Singapore: Springer Nature Singapore. https://doi.org/10.1007/978-981-99-3236-8_38

Pavlova, D., Dovramadjiev, T., Daskalov, D., Mirchev, N., Peev, I., Radeva, J., Dimova, R., Kavaldzhieva, K., Mrugalska, B., Szabo, G. & Kandioglou, A. (2024). Human factors and use of the surgical guide in dentistry – real practice example. In: Kaiser, M.S., Xie, J., Rathore, V.S. (eds) Intelligent Strategies for ICT. ICTCS 2023. Lecture Notes in Networks and Systems, 941. Singapore: Springer. https://doi.org/10.1007/978-981-97-1260-1_22

Shah, M. A., Lee, D. G., Lee, B. Y., & Hur, S. (2021). Classifications and applications of inkjet printing technology: A review. *IEEE Access*, 9, 140079–140102. https://doi.org/10.1109/ACCESS.2021.3119219.

Song, Y., Ghafari, Y., Asefnejad, A., & Toghraie, D. (2024). An overview of selective laser sintering 3D printing technology for biomedical and sports device applications: Processes, materials, and applications. *Optics & Laser Technology*, 171, 110459. https://doi.org/10.1016/j.optlastec.2023.110459

Tan, Y. J. N., Yong, W. P., Kochhar, J. S., Khanolkar, J., Yao, X., Sun, Y., Ao, C. K., & Soh, S. (2020). On-demand fully customizable drug tablets via 3D printing technology for personalized medicine. *Journal of Controlled Release*, 322, 42–52. https://doi.org/10.1016/j.jconrel.2020.02.046

Tren, S. J., Goyanes, A., Telford, R., Wilsdon, D., & Rowland, M. (2018). 3D printed drug products: Non-destructive dose verification using a rapid point-and-shoot approach.

International Journal of Pharmaceutics, 549(1–2), 283–292. https://doi.org/10.1016/j.ijpharm.2018.08.031

Trenfield, S. J., Xian Tan, H., Awad, A., Buanz, A., Gaisford, S., Basit, A. W., et al. (2019). Track-and-trace: Novel anti-counterfeit measures for 3D printed personalized drug products using smart material inks. *International Journal of Pharmaceutics*, 567, 118443. https://doi.org/10.1016/j.ijpharm.2019.06.034

Virmani, R., Kumar, G., Sharma, A., Virmani, T., & Pathak, K. (2024). Medical additive manufacturing in pharmacy. In *Medical Additive Manufacturing* (pp. 537–562). Amsterdam: Elsevier.

Waddington, L., & Priestley, M. (2021). A human rights approach to disability assessment. *Journal of International and Comparative Social Policy*, 37(1), 1–15. https://doi.org/10.1017/ics.2020.21

Wang, J., Zhang, Y., Aghda, N. H., Pillai, A. R., Thakkar, R., Nokhodchi, A., & Maniruzzaman, M. (2021). Emerging 3D printing technologies for drug delivery devices: Current status and future perspective. *Advanced drug delivery reviews*, 174, 294–316. https://doi.org/10.1016/j.addr.2021.04.019

World Health Organization (2017). A Study on the Public Health and Socioeconomic Impact of Substandard and Falsified Medical Products. Geneva, Switzerland: World Health Organization.

Yang, D. W. (2024). Disability discrimination in the provision of health insurance: Article 25 (e) of the UN convention on the rights of persons with disabilities. *International Journal of Discrimination and the Law*, *24*(1–2), 48–66. https://doi.org/10.1177/13582291241237987

Zapka, W. (2022). *Inkjet printing in industry: Materials, technologies, systems, and applications*. Hoboken: John Wiley & Sons.

Index

For Product Safety Concerns and Information please contact our EU representative GPSR@taylorandfrancis.com Taylor & Francis Verlag GmbH, Kaufingerstraße 24, 80331 München, Germany